About this Book

- A continuation of Book One.
- The main topic is mining exploration.
- The sub-topic is an individual's effort to make the transition from tents to civilization.
- Lots of interesting and sometimes amusing side-bits.
- Many locations herein: Timmins, Powassan/North Bay, Sudbury, Cochrane area, Fort Frances, Red Lake, Ignace, Kenora, Flin Flon, Pickle Lake and others.
- This could be anyone's story – we just happened to be the ones to write it.
- We pre-apologize – when you start reading you won't be able to put it down.
- Enjoyment guaranteed.

I Call Myself a Prospector

Book Two

Or

How to Find Your Way Out of the Bush without a Compass

By

Bob Durnin

With

Frank Durnin

Acknowledgements

AKA
Thank you list

To Carl Haglund and all those who read the first pitiful efforts, offering encouragement and useful input.

To Rick and Nicole who supplied me with a spiffy core shack to work in. – They kept the woodpile topped up and never once let me down.

To Ed Tetu, who believed in us.

To Bob Marvin, who had faith in a member of the Over-The-Hill Gang.

To Bell Canada & Virgin Mobile - 99% of the work was done with the two authors separated by 800 miles (1300k.)

To Bill Gates – couldn't have done it without ya.

To my co-writer – the true author – who gave me adjectives when I needed them.

To Janet, who stepped in when Little Brother got typist's cramp.

****

And to all those whose spirits still walk the Interlocking Olympic Rings

– our snowshoe trails will cross again someday.

Prologue

In Book One our anti-hero had literally (and often figuratively) gotten his feet wet in the mineral exploration milieu.

From 1965 onward is a tale of ambivalence. It is a case of addiction and withdrawal – two steps forward – one step back. Smokers know the drill. After three weeks they can throw the crutch away, but the psychological dependency is always there waiting to bite.

Civilization has its perks – clean sheets, automatic washers, refrigerators and people of the opposite gender to mingle with. Civilized people are nice to know and interesting to talk with. They have stories to tell also, but they mow their lawns, have regular days off, and sometimes when they listen to picket line stories you can tell they are thinking, "I wonder of this stuff is actually true?"

The weaning process is patchy. It starts with a season out, and the seasons become years. But the claim posts are always in the back of my mind, and the siren calls – she never rests.

So the story goes on, and the side stories go on, and the Interlocking Olympic Rings of the Canadian Exploration Fraternity dance across the Canadian Shield like the Northern Lights on a cold, clear winter's night.

Table of Contents

Chapter I   Back to the Diamond Drills –April – Dec '65

Summer in Powassan

The winter of 1965 had been an old-fashioned three-dog-nighter in January and February, but when the cold snowy winter's back was broken in March it became downright balmy. When we wrapped up the Mattagami job the snow was already disappearing rapidly.

I pulled out of Mattagami in the early morning of April 16 with no hard plan in mind other than to enjoy a month or more of relaxation. I do not yet know where the company will send me this summer, nor do I know when the summer season will kick off, but I am un-worried. (Whereas Inco had hard and fast rules regarding accumulated paid days off, Mining Corp. was more flexible. I would be on the payroll until the summer season started – at least one month away and possibly not until late May.)

I figure I will stop in to see my old friend Carl Haglund at Powassan, just south of North Bay.

Yup, good old Chattanooga Carl. Last summer after his return from North Carolina, Carl was looking for something to do. The area around his hometown was in a prolonged period of little rain. With many relatively shallow, hand-dug surface wells running low on water, Carl decided to go into the H2O exploration business. He bought a used but good Sullivan 12 from a North Bay exploration firm, found a '53 Fargo 4-ton with less than 20,000 miles on the clock, built a dandy enclosed van-type box on the truck, mounted the Sullivan inside and went drilling wells.

I pull into North Bay at Coutou Bros. Esso to gas up. I had met these boys back in '59 (on my 10,000 mile hitchhiking tour of North America) and since then have always made it a point to touch base. Butch comes out, sticks the nozzle in my gas tank and says, "Running a little rough, eh?" No "How are you doing?" or "Glad to see you," or such foolishness -he's worried about the car. Butch lifts the Oldsmobile's hood, scratches his head and goes to fetch his brother. Brother Paul comes out, listens to the motor and scratches <u>his</u> head.

"Shut her off," says Paul. He goes in and comes out with a new plug wire, which he installs. It took ten minutes and $15 to do a $400 valve job! Pay attention, people. If you find a good mechanic, keep him, it's like striking pay dirt.

I drive on down to Powassan and stop for a short visit with Carl's parents. They tell me Carl is on a well not far away, so I chug on over. Of course Carl is surprised to see me, but right now he is very busy pulling rods. The well is down 50 feet and the ten-foot core barrel must be pulled and emptied from time to time. Carl is on the drill running the winch, and his helper is on the ground. The drill rods are ten feet long and as Carl pulls the string it is the helper's job to set the rod brake after a ten-foot section has been pulled, break the rod coupling using two 24 inch pipe wrenches, help Carl unscrew the rod and stack it, screw the cable plug into the top of the string, release the brake and stand back while Carl winches the string up another ten feet. This process is repeated until the core barrel appears. The diamond bit and reaming shell are removed from the core barrel, and the core is dumped out. Then the whole process is reversed and drilling continues. It may sound confusing as I have described it, but it is not really a complicated operation. A trip out/trip in on a 60-foot hole should take no more than five minutes.

I can see that Carl has a problem. His helper is having one heck of a time breaking the rod couplings loose. Although the rods are screwed on hand tight when being lowered into the hole, the torque created when drilling tightens the couplings. There is a trick to breaking them free. You simply slap the wrenches in place, give them a good snap, drop the wrenches and help Carl spin the rod, stack it and continue.

The young fellow on the ground just doesn't have the knack of it. He puts the wrenches in the right place, but then he pulls and struggles instead of snapping them. It's almost as if he's trying to open a jar of pickles. Sometimes Carl has to climb down off the truck, break the rods and climb back to his station. It's slow going, and I can see Carl is kind of frustrated. They run another five feet and the core barrel blocks.

This time, I help Carl trip out and we are done in no time flat – the core barrel is out of the hole and emptied. It's 5 o'clock, so the drill is shut down for the day.

Carl and I retire to the local watering hole to have a cool one and catch up on the previous twelve months. The conversation comes around to the well drilling and the young fellow's inability to handle the pipe wrench concept. I half-jokingly suggest that Carl should hire me, but he doesn't think he can pay me enough. A couple of more bubblies follow, and true to the well-known fact that alcohol + $\circ$s = well thought out decisions, we hammer out a deal. I will work by the foot – no production, no pay. We raise a glass to the future. Onward, ever onward – I am back on the drills.

My room was second floor, third window back.

I take a room at the Windsor Arms and the next day I make a bed and breakfast deal with the owner at a reasonable weekly rate. I write a letter of resignation to Mining Corp. explaining my reason for quitting and they send me a letter thanking me for my services and a check covering four weeks of accumulated time off. A real class act, is Mining Corp!

Carl hates to let the young guy go, but has a talk with him, and he agrees that he isn't cut out for the job. By noon I have my ducks in a row and we finish the well and put the test pump on.

Here is the nitty gritty. We are drilling "A" core. "A" casing is 2" outside diameter and the core barrel is slightly smaller. In rock the hole bored is a little less than 1 7/8 inches, certainly not overly adequate for a well by today's standards, but in 1965 automatic washers were not all that common, and dishwashers were in the same category as UFOs. You know what I mean, "I've never seen one, but I know they are out there."

Our 1 7/8" hole in a good water seam can produce over five gallons a minute, and Ontario Water Resources (OWRC) mandates minimum acceptability at 2 gallons per minute.

Every time we pull the rods we plumb the hole and note the depth of water. We are using water under pressure to wash away the rock cuttings and when the rods are pulled, water remains in the hole. When we hit a water seam, the water running through the seam has its own static level, and the level in the hole will settle out at that depth. A dramatic change in the static level means we've hit water.

There are other indications as well. Getting into blocky rock is always encouraging, even though we may have to pull the string more often to clear the core barrel. Another indication is if the drill drops a little. Pressure exerted by the screw-feed always raises the truck up on its suspension a bit, and a quick drop of an inch or so means we have intersected a seam underground. Yet another indication is a change in the return water volume. If we hit a good seam we may lose all our return water, because the running water in the seam will maintain its static level, no matter how much we are pumping down the string. When we hit water we always drill another five feet or so to make a sump where rock chips and other detritus from the seam can collect. Now we have a well.

The next step is to pump it continuously for 24 hours to test its capacity. We put down a deep well injector, hook it to our test pump and throttle it until the water flow is uninterrupted. This tells us that the seam is replacing the water as fast as we are removing it, and now we can establish the well output. We measure the time it takes to fill a five-gallon pail with water. If it takes a minute, we have a five-gallon per minute well. It's not rocket science to figure out a lesser flow. A minimum of two gallons per minute takes two and a half minutes to fill the pail. A wristwatch and an old oil pail – this is hi-tech!

We now prepare the rig for transportation to the next well. It is an easy tear down. Carl has designed a simple tripod atop the truck box and we undo a couple of bolts and fold it down onto the roof. We stow the rods and core barrel on the floor of the box, throw in our gear – wrenches, tools, and lawn chairs (What do you expect us to sit on? You can't pull up a stump on someone's lawn, can you?) and leave the truck in place overnight, just in case we have to deepen the hole. The next morning we check the pump, and if the water is still flowing we move the truck to the next well, returning to remove the pump after the full 24-hour test – another successful well completed.

Our competition is an outfit from the Huntsville area. They are much more professional at the job, or at least you would get that impression if you compared equipment. They have a spiffy 6-cylinder diesel rig, built specifically for well drilling, mounted on a new tandem truck. It has all the bells and whistles and drills a 6" hole using an oil well type tri-cone bit. Carl has done an impressive job on his home built set-up, but if you put the rigs side by side his nice old '53 Fargo and wooden drill enclosure, no matter how neat and freshly painted, still looks like a country cousin.

However, we have a few things in our favour. Carl is local, born and raised in Powassan. His parents are honest, hard-working people from Sweden, and very community minded. Carl knows everybody, and everybody knows Carl. Carl charges four dollars a foot, the other guys six dollars, a pretty significant advantage in 1965. Also, our average well depth is not much more than 100 feet. Some are as shallow as 70 feet, but most are in the 90 foot to 120-foot range. We only have 210 feet of rod, and only once do we use the whole string. The Sullivan 12 designation means that it has a capability of 1200 feet, but that must be going downhill with a strong tail wind, because at anything over 150 feet our game little Hercules 4-banger needs frequent rest stops and radiator water transfusions.

The Huntsville outfit likes deep wells – 150 feet is peanuts to them. I'd never say they were unethical, but it makes you think. You can't set up a $40,000 rig, drill 70 feet and make money – it just doesn't compute. Now, here is our biggest advantage: The other guys have a history of 80% success – 20% of their wells fail to hit water – apparently acceptable by industry standards. We, on the other hand, never have a dry hole. Twenty-five or more wells that summer, and water in every one of them! Our secret? We have Fibber – they don't.

Mr. McGee is a little old retired railroader who lives in Powassan. I don't think I ever knew his first name and neither does anyone else it seems. He is Fibber McGee and he can witch water!

You may not believe in water witching, and neither did I before I met Fibber, but he could witch, let me tell you! Carl uses Fibber's talent on every hole, even if the customer is skeptical. Fibber never takes money. He believes cash might destroy his karma I guess, but after every well Carl gives him a mickey of Irish whiskey – a little whiskey never affects good karma. Carl told me that the year before he had only one dry hole and that was because the customer refused to believe Fibber, and wanted the well closer to the house. False economy, it turned out. From then on, Fibber's talents were used, whether the customer was a believer or not.

Fibber uses a green (as opposed to dead) willow switch. He finds one in the ditch, just the right size, with a fairly even fork. He then trims it to a "Y" with two-foot forks and a two-foot extension. Then he holds one fork in each hand with the end of the "Y" pointing forward. He then brings the end down and around up past his midriff, until it points forward again, making sure to keep a tight hold on the "Y" forks. Now his hands are upside down, with his thumbs on the outside, if you catch my drift. The flexible willow is bent fairly sharply at his hand-holds and the switch bobs a little as he walks. When he crosses a water seam the end of the willow switch points down: no water – it remains level. Fibber then does the same thing in sort of an X shape to make sure which way the seam is running. Then he will do a few parallel passes to check out the seam direction and we and/or the customer can pick our spot.

One day we play sort of a mean trick on Fibber. A guy had bought an acre to build on just outside of Powassan. He wants the well drilled before he builds, so he can situate his house near, but not over, the well. So we take Fibber out and he witches. The seam runs east-west, and Fibber is being careful to cover a large portion of the property when he does the parallel lines. He wants to make sure that he is not missing a stronger indication. We always put a stake in at Fibber's water location, so after two passes there are two stakes lined up east to west. As Fibber continues south, and before he turns north to do another pass, Carl and I quickly move the two stakes ten feet south. As Fibber returns north on his third pass, it seems to us that he is peeking a little sideways and when he lines up with those two bogusly planted pickets, sure enough – the willow switch dips again. Carl and I feel guilty as hell, and as Fibber continues north, we replace the pickets and move the third to line up with the others. On Fibber's fourth pass the willow switch dips on cue – right in line with the other pickets.

We drill, hit water at 120 feet, and it is a good well. Maybe this was just a broad seam, or maybe it was an area with a lot of seams; we'll never know. We also don't rat on Fibber, but our confidence in his witching is now a bit shaken.

Our next well is four miles out of town on the highway running to Chisholm (the place where the owner of the Ross House Hotel in Chattanooga once lived.) The road climbs a pretty good hill and at the top lives George Jones (I know what you're thinking, but I'm not goofy enough to make up a dumb name like that – his name really was George Jones) and George's well is going dry. He had built the house 15 years before, had dug a 15 foot surface well, and it has stopped producing.

We bring in Fibber. He witches the place and tells us there is a strong seam all the way up the hill towards George's house. Just below the house the seam splits, with a weaker indication to the existing dry well and a stronger one going around behind the house and beyond.

"OK," says George, "We will drill behind the house."

We set up the drill on Saturday. By 2 pm we have the truck in place and a half-mile of water line laid to a little creek at the bottom of the hill. We run twelve feet of casing down and collar the hole. On our first run we bring up a full core barrel. This looks like pretty easy going – the rock is not very hard. It's only 4 pm, but we shut down for the day and retire to our office chairs in the Royal Canadian Legion. Tomorrow is Sunday, and we don't drill on Sunday.

Monday morning George is gone to the bush, where he spends the week cutting timber. We fire up bright and early and pull another full core barrel. Wow! This has never happened before – two full barrels in a row! We run two more full barrels, and we are starting to get worried. To get water, you must have fractured rock, and even dry fractures mean that water may be near. When drilling, you always listen for a slight hiccup in the string, or even a slight jar may be felt as you rest your hand on the drill. This is something drillers do – listen and touch – it's second nature. But little Hercules and good old Sullivan sail along with nary a twitch, until a slight jar tells us that the core barrel is full again and the bit has cut the core.

Now we are really scratching our heads. Has Fibber led us astray? Is there really a water seam anywhere in this oh-so-seamless conglomerate? It's easy enough for Fibber to say the weaker seam runs in front of the house, after all he can see the dry well. It's far too late to quit now, so we run the string down again, but we are at 72 feet, have pulled six full barrels, and there is no indication of change in the static level in the hole.

Our bit is getting a little dull and the down pressure on the string is lifting the truck almost four inches on the springs. We figure we can get one more run out of the bit before changing it, and we keep a close eye on things. At 75 feet there is an audible "snap" from the string and the truck drops four inches - bango! The Hercules barks and settles down. Holy cow! Did we break a coupling? But no – the drill keeps on purring, and we continue our run. It's pretty obvious that we have hit something, because our drill water is no longer returning. We finish off the run, check the static level, and it's at eight feet!

We put the test pump on with 12 feet of suction line, pump for an hour, and the static level never changes. We've drilled 75 feet of seamless rock, and have hit the best well of the summer! Fibber is vindicated.

(If you need further proof of Fibber's powers, then read on. I'm sure that at the end of the next little tidbit you will become an honorary Fibber-Believer.)

We leave the test pump on for 24 hours, but we are so confident of this well that we pull the rig off that afternoon. We go on to the next job and George comes home on Friday, gets the good news from his wife, and looks us up to buy us a cool one at our Legion office. He strokes Carl a check, and is very, very happy to do so. He then leans back in his chair and tells us a story, and I don't mind telling you – a little chill runs down my spine.

"I didn't want to tell you this before you drilled," says George, "But I'll tell you now. When I built that house 15 years ago, Fibber didn't live here yet, but before I started building I had another guy witch a well for me."

"He traced that seam up the hill exactly like Fibber, and said the front seam, although weaker, was probably shallower. Just like Fibber, he said the better seam ran behind the house, but would probably require a rock capable drill. I dug the well in front to save money, and it has lasted me 15 years. That guy died 5 years before Fibber moved here. What do you think of that?"

Well I'll tell you, it is downright eerie.

Carl and I give Fibber's switch a shot. It doesn't work for Carl, but if Fibber walks beside me with his hand on my arm I seem to get results. The thing is – with the switch under tension, it doesn't take much more than a slight, undetectable change in your grip to make the pointer dip earthwards.

Sidebar: Twenty years later I would tell this story to a neighbour in rural Manitoba. He insisted we try it and I traced a water seam where his well was. He said I should hang out my shingle but I pointed out that I could see the well cap and my witching might have been bogus. And let's face it – these days, with well drillers charging 35 to 40 dollars a foot, a water witch has to have plenty of cojones to ask anyone to lay out that kind of cash on the advice of a willow switch – and besides, I don't like Irish whiskey.

Table Scrap

In 1990 my wife and I bought a country convenience store/gas bar not far from my boyhood home. In 1992 we added a restaurant/coffee shop and immediately ran into water problems. In 1990 we had dug an 18 foot surface well with a four foot crib, but the recovery rate was unable to handle the demand for the expanded enterprise. I had to haul 2000 gallons each week from a town six miles away and it was a real pain in the butt - we needed a drilled well for sure. The problem was money. The well on an adjoining lot went down 190 feet, and at a house 200 feet away, the well was over 200 feet deep. We were already into the red due to the expansion, and we couldn't justify a further expenditure of $10,000 that year.

One of our regular customers was Mel Jack, the local well driller and a long-time friend. He was a dandy driller with good equipment. He was also known for his firm insistence for cash on the line. When Mel pulled the test pump, he expected to leave with the pump in his truck and a check in his pocket. It was just business, that's all, and I never heard anyone bark about it. When Mel drilled you paid – pure and simple.

So Mel drops in for coffee, sees the water truck beside my well crib, and asks what the deal is. I tell him, and Mel says. "Drill a well."

I tell him maybe next year, we can't afford one right now. This goes on for a month or more. Mel says "Drill a well," and I say, "Next year."

Now, Mel knows how hard my wife and I are working to get this business on its feet. We open at 6 am, close at 10 pm, and we are always here – always! So one day Mel comes in and sits down with me at the coffee table. There is no one else here at the moment. Mel looks around and turns to me.

"Bob," he says, "I am going to say something to you right now that I have never said to anyone in my life, nor will I ever say it again, and if YOU ever repeat it, you can say goodbye to old Mel 'cause we will no longer be friends."

"I'll drill you a well," says Mel, "And you can pay me when you have the money!!"

I am in shock – I never expected this, and I damned near have tears in my eyes. We go out to Mel's truck and he gets his witching gear. He uses two copper welding rods to witch. Each rod is three feet long and bent 90 degrees, four or five inches from the end. Mel holds the rods pointed out straight ahead of him as he makes his passes. "Where do you want the well?" he says.

We are at the northeast corner of the restaurant, and I say "Anywhere back here is just fine."

Mel makes a few passes and the rods cross at a spot about 15 feet east of the corner of the building, and maybe five feet in from the lot line/highway right-of-way.

"There's your well," says Mel.

"Pretty close to the line," says I. (Bell Canada had recently buried a fibre optic cable along the highway and the word was out that if you cut the line, Bell would charge you a ridiculous amount of money for each minute of fibre optic interruption.) We eyeball the lot line and it looks like we're OK.

So the next morning Mel's driller comes in, and by 7:30 am he starts drilling. I don't hang around. I have other things to do, and I expect him to be there a couple of days, anyhow.

At 9:30 am I am unloading a beer shipment into the cooler on the west side of the restaurant when Corey, the driller, comes around the corner.

"I've got good news and bad news," He says, "Which do you want first?"

My heart skips a beat or two. I'm sure we've cut the fibre optic cable. "You might as well give me the good news first," says I.

"The good news is that we hit water at 37 feet," says Corey.

"Holy cow," says I, "So what's the bad news?"

"The bad news is, I'm not making any money!" he says.

We need a sump anyhow, so I tell Corey to run her down another ten feet.

By 11 am the drill is gone and the test pump is spewing out five gallons per minute. At 11:05 the next day Mel has a check in his pocket, the backhoe comes in, and by 4 pm we have the water line into the basement and the restaurant has an unlimited supply.

I feel like we have hit the lottery jackpot!

Mel is still drilling wells and we are still friends. I'm sorry to rat you out, Mel, but I checked it out. The statute of limitations on the "Mum's the word" law expires after 20 years.

Mel tells me he doesn't witch anymore. I guess Mel has lost his karma.

Carl and I move to our next well and run into a spot of trouble. Fibber doesn't have to witch this one - the customer has a 40 foot well with six inch casing, and because the well is already in rock, he wants us to deepen it. His water flow has decreased due to the drought and it probably wasn't that good a well to begin with. If he calls in the big rig they will have to start from scratch and he wants to save a buck.

We don't like to go down an existing well for a few good reasons. For one thing, we can't charge enough for the first 40 feet to make any money. We have to put the casing down and collar the hole before we drill, and although we can retrieve the casing and diamond casing shoe at the completion of the well, it's hard to recover our cost of set-up and tear-down time. And on top of all this, I won't get paid for the first 40 feet.

The second thing is that the drill must be plumbed exactly in the centre of the six-inch casing. The slope of the back yard and the tight quarters make the setup interesting as well. We don't want to damage any of the customer's mature evergreens or ornamental bushes.

We manoeuvre the truck into place, make an educated, highly technical eyeball-guess, borrow a couple of railway jacks and a stash of 4"x 4" – 4 foot timbers, and block the front wheels up more than three feet to level the drill platform. The next day we collar the casing, drill 30 feet and find water with a static level of 15 feet. The customer is very happy. He has a well now, with a 25-foot, six-inch diameter reservoir for $125.00 plus a few peanuts. We have spent two days and a lot of hard work for the money. We would have been better off working in Berzerkerstan for rancid yak butter.

We could have put the test pump on and gone home for the day, but oh no, not us! We still have a few hours left in the day to shoot ourselves in the foot. Whether it is that we are a little cocky because of our success at George's well, or because we are eager to get on with another better paying well to recoup our losses, we pull the drill off before testing the well. We take out the blocks, lower the truck, and while Carl moves the truck away I police the area. I'm clumsy as a mule anyway, and as I pick up the core lying around the well I drop a five-inch piece down the damned hole.

"Holy cow! Maybe I'm lucky," I think, "Maybe the core went down to the bottom of the sump."

No way baby, we can't get the test pump injector past the 40' mark. The core is obviously crossways at the bottom of the six-inch casing.

Maybe we can break the core. We haul fifty feet of rod back to the hole and piece-by-piece, we lower them by hand, and holding on very tightly, we try to hammer the piece of core. If we are successful, we don't want to lose our grip on the rods. If they drop into the 70-foot well, we will have to add another 30 feet of steel to get them back and it would take our drill engine's namesake to handle the task. (No worries about losing the rods – we are unable to break through.)

We consider our options. We can say nothing and hope the customer doesn't find out – it is possible that an injector at 40 feet will suffice, but we have our ethics, so that's out.

Carl goes to North Bay and buys a core grabber. It's not an expensive deal, just a short fitting that screws onto the end of the rods, and with a little spring dealie inside. The theory is simple. The core grabber fits over the end of the core and the spring holds it as we pull the rods. It's a long shot, and it doesn't work. The core has to be on its end for the core grabber to have something to hold on to, and we already know it's crossways. We are grabbing at straws, unwilling to face reality. Finally, we bite the bullet. We have to put the drill back on the hole.

It takes even longer to line it up the second time. Finally on our third try we get the truck in place, raise our tripod and send the core barrel back down the hole. Two minutes of drilling and we are through the piece of core. We pull the string and the remaining pieces of the core fall into the sump. We leave the drill in place, put the test pump in and beetle off to our office to anoint our wounds with liquid barley.

Just to prove we are consistent, we fire the other barrel into our other foot on the next well. We are set up on the outskirts of Powassan and we collar the well at 30 feet by 11:30 am. We feel pretty good – this could be a one-day well. We run the core barrel down and drill three feet to make sure we're in bedrock, then shut down and head for the office for a noon sandwich, leaving the core barrel in the hole. After lunch we will run the barrel full and continue with the well.

One sandwich leads to another and it's 2 pm when we get back to the drill. We fire up the pressure pump and the Hercules, and Carl pulls the clutch – and stalls the motor! He tries one more time but the drill string won't budge, and to top it off we don't have any return water. I put a pipe wrench on the rods, slide a five foot length of casing over the handle and pull. The string turns a little bit, but so does the well casing. We are sanded in!

Here is what happened. When we drilled that three feet before lunch we had intersected a small water seam filled with fine sand. Had we been paying attention, this would have not been a big problem. Our drill water would wash the sand to the surface, or help carry it away with the underground flow. If the sand problem had persisted, we would have simply run the casing down another three feet and the seam would have been sealed off. However, during our extended lunch break the sand had percolated up and filled the small space between the inner wall of the casing and the outer wall of the core barrel. The two are now essentially welded together. With much difficulty we pull the drill string and casing simultaneously, unscrewing each ten-foot of casing and rod as they are pulled out of the hole.

We are able to remove the diamond drill bit and reaming shell, but the core barrel is trash. It is stuck permanently inside the last ten feet of casing. After two hours of separation anxiety we give up. The loss of the casing is no big deal, but he core barrel is tool steel, and is a pricey item.

There is an old adage that says. "The only people who never make a mistake are those who do nothing." Well, I disagree. Two hours of doing nothing was a big mistake, that day.

(In retrospect I think the Royal Canadian Legion should have bought us a new core barrel. Had they opened the bar at noon instead of 11 am we would have finished the first run.)

In mid-August the rains come – the drought has ended. Wells are filling with water and the well drilling is drying up, proving that for every action there is an equal and opposite reaction. We have one last well ahead of us.

We are prepared for the downturn. When the rains started, Carl had also started putting out traps on his line. He has made a deal with J. K. Smit on a new Winkie drill and associated gear. We also have three possible contracts. One is a referral from Fred Wink. A small group of wannabe promoters from Tonnawanda, New York, want two holes on their mining claims at Lake Abitibi, east of Cochrane.

We have two other jobs lined up, one in the Hemlo/Manitouwadge area. Consolidated Mining and Smelting (Cominco) of Trail, B.C. wants 1200 feet drilled near the Little Black River, half-way between Manitouwadge and Highway 17 (Trans-Canada Southern Route.) Carl has also talked to a prospector/promoter friend of his in Toronto, who wants some drilling done on a property near Rouyn, P.Q.

We are back in the bush! The question is, which job do we do first?

Little Black River and Beyond

We talk to the New York people first. We had started our last well the day before and had collared the casing. Today we meet the Tonnawanda group in the Windsor Arms dining room for breakfast. There are four of them. A couple of guys are over 60, I'd say, and the other two are a bit younger. All are well dressed and none of them appear to be particularly goofy. Their spokesman is a blustery type and the other three are head-nodders.

We start the first slow dance. They tell us where they're coming from, we tell them where we've been. We try not to lay it on too thick and assume they are doing the same. We start to fox-trot. How much will the 800 feet of drilling cost them? That depends on where it is, exactly, and how we get there. They tell us the claims are on the north side of Lake Abitibi, 30 miles from the road access on the south shore, and they will supply boat transportation. A Frankie Yankovich Polka is put on the turntable. What kind of boat? They have a 32-foot sailboat! Carl and I miss a few steps – a sailboat on Lake Abitibi? Don't they know that Abitibi is a shallow lake? And with any kind of wind the waves are tremendous? We also point out that with camp gear, drill, over 400 feet of rods and other assorted gear rolling around on deck, we really can't see ourselves heeling over on a windward tack. Frankie is really giving his squeezebox a workout now. No problem, they say, we don't need the mast and sail, the boat has a 15 horse outboard and as for the rough water, a jury rigged wall of plywood should handle that.

Carl and I are almost danced out, but out of curiosity we try one more, and this go-around can only be called wildly interpretive. How do we get paid? No problem. (Nothing is a problem for these boys.) They will stay with us until we finish the first 400-foot hole, then they will leave us to drill the second hole while they take the first gold-bearing core back to Tonnawanda, where they will raise oodles of money. They will then return to their gold mine on Lake Abitibi, retrieve us and the rest of the core, and lay the cash into our hot little hands. Cymbal crash – the dance is over.

They want to watch us drill, so they follow on the 20-mile trip to the well. On the way Carl and I are having a good chuckle. We check each other's foreheads for "stupid" tattoos – nope, just zits.

I guess they figure that if we believed that first song and dance we would believe anything. They ask us how we decide where to drill, and we tell them about Fibber. The head crazy person goes to his car and brings back something resembling a big spider missing some legs.

"I call this the Bob-o-Link and I can witch water with it."

The Bob-o-Link looks like a space age version of Fibber's willow switch. Instead of willow, the "Y" is made of high-tensile springy steel rod. At each end of the Y are aluminium canisters shaped like bratwurst sausages. He holds the thing out in front of him, and when he gets to the well, the contraption starts bobbing up and down. Each bob means five feet, and he counts 13 bobs.

"You'll hit water at 65 feet," he says. Then he turns it upside down and it bobs three more times. "Your static level will be at 15."

Carl and I are dumbfounded. (Dumbfounded: adverb – to be found dumb as a sack of hammers by four Tonnawanda goofballs.) We tell them we will get back to them and we say goodbye. We already know what our answer is, "Thanks, but no thanks."

(Incidentally, we hit water at 135 feet, and it was a flowing well – the static level was five feet above ground.)

But – as we write this there is a huge open pit gold mine in production on the north-west shore of Lake Abitibi – maybe Tonnawanda wasn't so crazy after all.

So now we are down to two choices – Cominco at Little Black River, or Rouyn. We chose to do the Cominco contract first.

We put the Sullivan away for the winter and Carl heads to Toronto to pick up the Winkie. I settle up with the Windsor and pick up my gear. Most of my stuff is stored in the Oldsmobile's trunk and I will leave the car in Powassan for now.

When Carl returned from North Carolina last year he had traded the Bel-Air for a new '64 GMC six cylinder pickup. It is a good economical rig and quite adequate for our purposes.

Carl returns with the drill and we load the truck with camp gear, sleeping bags, packsacks and tools necessary for a bush drill job. We need a third man, so a local guy is hired to go with us. Dennis is married and works for a local garage. He is a pretty good guy and we have been extolling the romantic adventures on the Canadian Shield. He decides to join us, and with his boss's blessing and assurance that the garage job will always be available, he throws his hat in the ring.

(Actually, this will not be Dennis' first rodeo. In their late teens he and Carl had spent the summer prospecting for Hasaga Gold Mines. He already knows how romantic black flies can be but I guess our enthusiasm is catching.)

The day after Labour Day we hit the road early. The GMC is loaded to the hilt, and with three galoots in the cab we head off into the wilderness of the Lake Superior North Shore. We have a ten hour drive ahead of us. With the load we are carrying, it doesn't pay to be heavy on the throttle.

I am familiar with the Little Black River area. It's not too far east of Marathon and halfway between Hwy 17 and Manitouwadge to the north. (In '61 I worked out of Marathon and had spent a couple of months at Manitouwadge in '64.) We plan to overnight at Heron Bay, a little village not far from Marathon. A Cominco geologist will meet us at the Little Black River tomorrow morning.

We are tooling along and it's a great day. We pass Wawa and at White River we stop and pick up a six pack of Labatt's Blue in cans, the idea being that we will have a couple of cool ones before retiring early. Tomorrow will be a busy day. We pass the Manitouwadge junction and it's not far to Heron Bay. What the heck – we crack a beer and sip a suds.

Back between Wawa and White River we had met a transport and the gust of wind had blown our flimsy mirror off the door. No problem – we have no rear vision due to our large load, but tomorrow the truck will be empty and the windshield mounted mirror will suffice. We finish our beer and turn off on the five mile road to Heron Bay. I am driving, and since we don't want to be carrying empty cans in the cab, I roll down the window and fling out the empties. Holy cow! I hear a siren and a black car with a white door runs up alongside. Busted!

I pull over and a young officer walks up to the truck. Just as he reaches our door the horn starts blowing and it won't stop! I hammer on the horn button, to no avail. I finally jump out, lift the hood and yank the wire off the horn relay. I go back to the door where the cop is talking to Carl and Dennis.

"You boys have any beer in that truck?"

"Oh, no, not us," we say, our angel faces on.

"That's odd," he says, "Because an empty can just bounced off the hood of my cruiser back there."

We fess up and explain that we had each nursed a beer over the last 50 miles. We show him the six-pack with three empty spaces, and he accepts that part of the story.

"You'd better follow me to the detachment," he says, so we go to Marathon.

At the detachment, I get the impression I will be spared from death by firing squad – he's told the other cops the deal, and they all think it's kind of funny. I pay a $20 fine for "liquor other than at a place of residence." and we are on our way. Not an auspicious start on this contract, but it could have been worse. The cop could have thrown a whole list of charges at us. We were littering a public highway and driving an unsafe vehicle (no serviceable rear view mirror) just to name a couple. I am happy to pay the 20 bucks.

Sidebar: The law in Ontario was (and maybe still is) that if you purchased alcohol you were to transport it to your residence by the most direct route. Theoretically, you broke the law if you stopped to buy mix on the way home. If you are a registered guest of a hotel/motel, then that is your place of residence. On that day, at that time, we had not yet registered at the Heron Bay Hotel. We were "of no fixed address."

We check into the hotel and have a quick bite to eat. It's Saturday and we have to pick up some groceries before the store closes, so we go next door to the store. That's pretty much it for Heron Bay, one short main street with the Canadian Pacific tracks on the south side and the Heron Bay Hotel and store on the north. Behind the hotel and store is a short street with three or four houses.

We walk into K.T.McCuaig and Son, and we are in a time warp. Here is everything from magazines to ammo, with dry goods, hardware and groceries in between; if KT doesn't have what you need today he can get it tomorrow. Need a taxi to the dock? KT and Son can supply it. Need a log boom delivered from the mouth of the river to Marathon Pulp & Paper? KT and Son can tow it. Look up the definition of "broad-based business enterprise" in the dictionary and you will see KT's picture. I've heard of KT – who hasn't? The man fits the legend – he's quite a character.

We buy enough groceries for a week. It's a two hour drive to our proposed campsite, so weekly supply trips are feasible, and besides, we won't be here that long anyway (take note of this statement.) We also need a flashlight. I pick up one of those six-volt dealies with a handle and a large lens. "Is this rugged enough for the bush?" I ask.

KT picks up the flashlight and fires it the forty foot width of the store, where it bounces off the far wall. I pick it up and it still works. "I guess that'll do you," says KT.

We put the fresh meat on ice in our cooler, load the grub, give KT a cheque and head back to the hotel. I said we were going to hit the sack early, right? Wrong. We close the bar.

Sidebar: Every old mining town I have ever been in has the equivalent of a K.T. McCuaig and Son. They know what you need and know how to pack stuff for the bush. Red lake, Timmins and Flin Flon come to mind. Heron Bay and Savant Lake also, although not really mining towns, are frontier towns, and when you shop in these establishments you don't shop prices – you shop service. In those days everyone would take a cheque, and you don't dare mess with your credit rating. Bounce a cheque in Heron Bay today, drive to Red lake tomorrow, and sure as hell the moccasin telegraph will get there before you.

The next morning Carl and I are up before dawn and haul our stuff down to the pick-up. We hammer on Dennis' door on the way by and I think I hear him answer. We stand by the truck and wait – no Dennis. We go back to the door and the snap lock is engaged – we can't get back in. What to do? It's 6 am Sunday morning and the hotel has already told us the dining room is closed Sundays.

No problem – the Heron Bay Hotel is built to the same floor plan as so many of its era. The rooms run down each side of the second floor, and by counting windows we can figure out which one is Dennis'.

We try throwing rocks, but they just bounce off the screen. I walk around to the back of the hotel and find an extension ladder. I climb up to the window, remove the screen and raise the sash. "Wakey, wakey, it's daylight in the swamp!" I holler.

Dennis comes down ten minutes later, rubbing his eyes. In fact, he comes down the ladder. I insist he do so. Dennis thinks we are nuts – he is, after all, a town guy. He has yet to learn the code of the bushwhacker – "Drink all night, never get drunk. Work all day, never get tired." We just have to educate him, that's all.

The next time we go to Heron Bay for groceries, the hotel owner thanks us for taking down the screen. But he's a little pissed that we didn't put up the storm windows or put away the ladder – Eh, boys?

We meet the Cominco geologist 15 miles up the Manitouwadge Highway, and follow him in on an old gravelled trail to the Little Black River. Here we find a cleared area the size of your local supermarket parking lot. The wood-cutting camp is long gone, and the Cominco guy says we can camp here. We don't like it – it's too close to civilization for us and we want our camp out of sight. We explore a little-used trail that leads off down the river and find a nice level grassy spot on a low part of the river bank. There's lots of room for one tent and one pickup truck, and it's out of the way of prying eyes and sticky fingers. We soon have a good camp set up and make ourselves at home-sweet-temporary-home.

Table Scrap

In the centre of the supermarket parking lot there was one log building, surrounded by smooth gravel. It was built like a blockhouse, with no windows and one door.

The low-slope roof showed no shingling. In fact it looked like an old garden with a heavy matting of dead grass and weeds. My mates (such children) wanted to explore the possibility of using the building rather than pitching our tent. I did not agree and did not even drive very close to it on our way to the river, and here's why.

Back in the day, timber camps often built these structures for meat/produce storage. The walls were made of heavy logs and after the roofing was nailed down they added a foot or more of dirt as an insulation factor. Thus it was cool in the summer and frost-resistant in the winter.

Thanks to Gerry, who I had worked with last fall, I knew we would keep our distance. We would <u>not</u> live there – neither for three weeks nor for one night.

Gerry had told me that a few years before they had set up in such a place. I have pointed out in Book One that any kind of a shack is a step up from a tent, and Gerry had thought likewise. His story gave me shivers – nightmares, even.

They set up housekeeping, fired up the airtight and settled in for a comfy night. It was late fall – not much snow yet but already well below freezing. In the middle of the night they heard some noises overhead. A quick check with a flashlight shone on hundreds (?) of snakes in the rafters, awakened from a winter's hibernation in the sod roof!

Women and children first? No way – the door was not wide enough for the three-man stampede!

They slept in the truck and the tent went up the next day, far, far away from the blockhouse.

Cracks in our Cominco contract are evident from day one. When we arrive at the campsite we have a few questions for the Cominco guy, and information is not that easy to yank out of him. He is sort of a tight-ass, not too smiley and all business. It turns out he is new to the job, and either is a natural-born jerk or, more likely, trying to project himself as a hard-nosed, not-to-be-trifled-with future project manager. I think he is a recent graduate with a fresh diploma on his wall, but he does not yet have an attitude adjustment certificate.

Boat transportation is to be supplied – where is the boat? The boat will be here tomorrow morning. Core boxes are to be supplied, where are they? Core boxes will be sent. At least we were smart enough to anticipate this. We have brought enough boxes for the first hole and the geologist promises to replace them. (They never do so.) The guy tells us that the prospectors will bring our river transport the next morning and they will show us the three hole locations. Without saying goodbye, he disappears. He is heading back to Thunder Bay.

The next day two young fellows arrive from Manitouwadge with our river transportation – a 17 foot v-stern cedar strip canoe with a three-horse kicker. They take us into the claim block where the three holes have been spotted, and it is abundantly obvious why the young geologist had beetled off so quickly. If he had been handy we would have thrown him in the Little Black and sat on him until the bubbles subsided.

It was really our fault when you come right down to it. We had read the contract and Carl had signed it, but we are so naive we shouldn't be allowed on the streets without our mommies. Here's the deal.

The job was to be a half mile down the river by boat and a half mile in the bush, with water available at the drill site. It's half a mile down river all right, then a 1200 foot portage around some rapids and on for another half mile by 12 foot skiff. It's a half mile into the bush, too, if you don't count the first mile. As for water availability, there is a little boulder-filled depression with a few puddles of water – enough to wet a squirrel's whistle. We are semi-screwed. We should have written mobilization/demobilization into the contract, and we should have written in cost-plus to cover any discrepancies between stated facts and actual on-the-ground facts. The portage and extra river and bush travel should have been at cost plus 10% in and out every day. Our three holes of 400 feet each turn out to be one hole of 300 feet, one hole of 400 feet and one of 500! The extra length is within the Winkie's depth range, but we had bid the holes on the basis of 400 feet each. The extra hundred feet should have been at cost plus, or at least at a one dollar per foot premium. We had trusted Cominco to be fair with us and we can't back out now. Rouyn is looking pretty good, but we have shot our company bankroll and it's too late to say "Whoa!" in this particular mud hole. We bite another bullet and get to work.

Hauling the Winkie around in a tippy canoe is not easy. You don't dare drop anything – cedar strip canoes are not all that rugged, so many careful trips later we have all our stuff at the head end of the portage. Of course we have parcelled out the work – while two guys haul the gear by canoe the third starts portaging. Then – with everything at the bottom end of the portage, we bring down the kicker and continue on by boat – if you can call a 12 foot skiff a boat. It's a little more stable than the canoe, but with two men there is not much freeboard for drill, rods, etc. Once again, one man remains at the last landing and packs stuff into the drill site.

There is one shining light in the whole dark mess. Cominco had optioned the ground from the two young men. They are brothers, recent immigrants from Yugoslavia – Croatian I think – and work in the mine at Manitouwadge. They have whole-heartedly taken up weekend prospecting and, naturally, have a vested interest in the program. They are lanky, energetic, and always have a positive attitude. They come down on their days off and help us, and always refuse to accept any remuneration for their efforts.

(Carl and I were talking about this just a day ago. We regret that we didn't stay in touch with those boys. We can't even remember their names.)

First we tackle the water problem. We don't have a mile and half of water line nor do we have the cash for the auxiliary pressure pump that length of line requires. We do a quick set-up at the so-called water hole, drive down ten feet of casing, move the drill, drop in three sticks of dynamite (supplied by the prospectors,) pull the casing, light the fuse and head for the bush. Result? A decent-sized water hole – good enough.

We collar the first hole in bedrock at 20 feet. It is a tough go through a boulder bed, and the next two holes are likewise. We had been told that the overburden would be no more than five feet deep, so of course, we didn't specify cost-plus, or a higher rate for heavier overburden.

We _were_ smart enough to bring 50 extra feet of casing and two extra shoes (bits) with us. Even on a hole with little overburden it's not always possible to retrieve the casing.

22

And we do lose some casing and a shoe on one hole which we had collared at 25 feet. The shoe stubbornly refuses to let go. To retrieve what we can, we use our pipe wrenches to unscrew the casing – if we are lucky it will leave only the shoe behind. No go. We recover only five feet and leave $20 worth of pipe and a $50 diamond encrusted casing shoe in the depths of the shield.

We wanted to be out of here before the non-resident moose season opened, but with the long trek, the difficult overburden and Cominco's less-than-honest description of the contract, we are still on the job on September 30.

I had been in the area last moose season, but even I am shocked. One day the old bush camp parking lot is empty, the next day you can't find room to park a dirt bike. Slide-ins – tents – motor homes – you name it – it looks like a small town. The season opens tomorrow and we don't have one lick of hunter orange – just our red life jackets – so we put them on.

The next morning as we walk down the portage we sing at the top of our voices and hope we don't sound like a moose. (The year before a hunter had put two rounds through the cab of a road grader. He said he had a good "sound shot!") We climb in our punt and head down river. The river bank south of the portage is quite high on the east side with a treeless bank of clay topped by a few scrub pines and bushes. On the west bank there is a low swamp with small willows, canary grass, and some open pools of shallow bog water – ideal moose feed.

As we push off, Dennis, in the bow, points up the river bank. He doesn't say a word – just points. We look up, and holy cow! The ridge is loaded with hunters – at least fifteen, maybe more, all with guns pointing towards the swamp, and we are in the line of fire. We holler "Don't shoot!" and move on down the river as fast as the kicker can go. We piss them off for sure, but we just want to arrive alive.

(I've since wondered what would have happened had a moose appeared on the other side. With 15 or 20 rifles on the ridge, how could they know who shot the moose? Maybe they would spit the carcass 20 ways, one piece of meat per bullet hole.)

We are almost finished, so we pour on the coal and four days later we are hauling stuff out. It's remarkable how fast we move, no doubt due to the incentive of clean sheets back home. The two young prospectors are a great help, and by evening we have most of the drill loaded in the pickup. A half-day of portaging remains, so the boys shake hands and say goodbye – they have to work tomorrow. We will stash the boat, canoe and motor for them to pick up later. We are all a little disappointed that the core didn't look that good but we point out that none of us are geologists. Maybe the professional core log will prove to be more interesting.

By noon the next day we have the gear out of the bush and on the pick-up. Carl and Dennis start to pull the tent down. I get into the truck and the battery is dead, stone dead, not even a click – and I know why. On our last trip to Heron Bay I had checked the oil before leaving camp. I saw the horn relay wire hanging there and plugged it in. The horn button worked, so I left the wire on. Well, mechanical things don't heal themselves, and I guess the horn decided to blow one day when we were in the bush. It blew and blew until the battery died.

This is a bummer – but no problem. I'll just walk up to the old camp and find a hunter who is willing to give us a boost. I walk up to the hunter village, and it's empty! Not a vehicle in sight! Four of five days ago this place looked like an RV trade show. I can't believe it! I go back and tell my partners the bad news.

We realize what has happened: we are the cause of the exodus. After opening day we saw no hunters on the ridge and had paid no heed to that. Our little Winkie could be heard for five miles away, and when we started drilling they must have gone someplace quieter to hunt. We certainly didn't see any moose – I guess we had already chased them away before the season opened. I imagine a horn blowing for two hours didn't help either. We are alone on the river and my name is Mud. We are like atheists in a funeral parlor – all dressed up, and no place to go.

I have a brain wave. We have a hundred feet of 5/8 inch rope that we use to pull rods on deeper holes. (Even though we use lightweight aluminium/magnesium drill rods, 400 to 500 feet is a heavy lift, so we rig up a snatch block on a tripod to pull the extra weight.)

I figure we can jack the truck up, wrap the rope around one rear tire, put the truck in gear, grab the rope and run like hell. After a few false starts we get one good run going, but the engine won't fire. In 1964 all vehicles had a generator. Alternators were introduced by Ford in '65 with other manufacturers following suit in '66. Alternators put out more juice at low rpm but generators need to spin faster, and with our battery stone cold dead, we can't run fast enough to generate a spark at the plugs.

We give it one more try and give up. I think it's kind of funny – imagine three yahoos running with a rope leading back to a jacked up truck on a riverbank in the Canadian wilderness. Carl and Dennis fail to see the humour in the situation.

So, who is elected to walk out five miles to the highway? The horn relay guy, that's who. It's Saturday afternoon and it may be tough to find someone willing to come in to assist us. I may have to hitchhike into Manitouwadge. Carl and Dennis will wait until 4 pm and if I'm not back, they will set up the tent again.

Just as I am about to leave, we hear an outboard motor and, by golly, a little boat is coming down the river! It's a hunter! We are saved!

We rush down to the river to intercept him and he is obviously startled. He pulls over to the far bank and asks us want we want. We tell him our predicament and he offers to go back to his vehicle and bring his battery and jumper cables. Soon we are mobile again. The guy heads quickly back up-river. We barely have time to thank him and he leaves before we can offer payment.

I know why he is so eager to depart. When we first hailed him, he had stopped on the far bank and thrown his jacket over something in the boat. As I helped him with the battery, I saw a tail sticking out. The bugger had shot a beaver! We must have scared the dickens out of him when we hollered – he figured he was busted! I'm sure it wasn't his trap line, and beaver will not be prime for a month at least. He was one of those guys who has to shoot anything he sees, I guess. He has done us a favour, so we are not about to rat him out.

We finish loading the tent and our personal stuff and leave at 4 pm. We arrive at Powassan in the wee hours – Little Black River is behind us.

The next day Carl calls Cominco to tell them we are finished. The prospectors have the core, and the final invoice will be mailed to-day. This time, Carl talks to the head geologist and tells him how disappointed we are with them. The guy can see Carl's point of view, and is quite commiserative. He agrees the contract was unfair, and Cominco will pay for demobilization. It's not much, but at least Carl can come close to breaking even.

Sidebar: We had chosen to do the Cominco job first because it seemed to be so straightforward – three four hundred foot holes with a half-mile walk should take 21 days max, pin-to-pin.

Our bid was simple – and so were we – four dollars a foot all inclusive. The bid should have been more comprehensive, and by leaving our fate in the hands of the "opposition," we left almost $2000 on the table.

We had bit the bullet thinking that we would get subsequent work from Cominco. They never called us again.

(Carl went on to the Rouyn contract alone, using local labour. The Toronto syndicate, with shallow pockets, treated him well, making sure Carl made a buck – a much better business model.)

Table Scrap: Further to putting one's trust in large corporations.

In late winter (early April) of 1970 I was nearing the end of my tenure with Noranda. They had some claims on Swell Bay twenty miles east of Fort Frances. I ran a quick survey on the lake ice and picked up a conductor. Noranda wanted to put down a quick 400-foot hole before the ice melted.

Sam Duggan, their usual contract driller was busy elsewhere so they gave the job to Art D, a local driller. Art brought in his BBS-1, set up on the lake and started to run casing. He hit lake bottom at 80 feet and continued on with the casing – and on – and on.

He was drilling at 65 degrees and encountering silty clay – not a rock or boulder to impede his progress. By the time he had 150 feet of casing down he was running into problems. With no firm overburden, the string was drooping and he was in danger of breaking the pipe at depth. He pulled back and fished some heavy rope through a hole in the ice to snag the string at 80 feet and provide some support. At 200 feet he was still in silt and the casing was getting hard to manage. He had bid the job at X dollars per foot of casing and was losing his shirt.

I had a chat with Art – telling him he had to talk to Noranda – this should have been at cost-plus.

Art said <u>he</u> would bite the bullet in order to get more work. I told him that when Sam came back they would never call him (Art) again. They never did.

I had my doubts about the whole deal anyway. My interpretation of my survey was that they were chasing conductive lake bottom (which they were.) It was just another reason I was severing my ties with Noranda. Why pay me and ignore my advice?

Carl and I part company. I feel I am abandoning him at a bad time, but there is not enough money in Winkie contracting to support two of us, and I will soon be getting married. Thanks to Carl and his North Bay contacts, I have a job drilling grout holes at a hydro dam under construction near Renfrew in the Ottawa valley. Carl is going to Rouyn to drill for his old syndicate friend and Dennis is going back to being a mechanic – he has tried the bush life and it's not for him. Dennis is also sure we are a bit mentally unbalanced – and he's probably right.

I am off to the Mountain Chute dam the next day. I need a place to stay, so I go to the little town of Dacre, not too far from the new job, to look up young O'Reilly, who I had befriended on Nighthawk Lake last winter. He is away working somewhere, but his mom and dad are very happy to see me. Their son has told them how helpful I was on his first foray into the bush. I ask them about a place to stay and I am told that they would be disappointed if I didn't stay with them. Cool – I have room and board within 25 miles of work.

At 8 am the next day I find the project and sign up. A young French guy from Quebec hires on at the same time. The construction company needs two drillers, one will be running a Winkie and the other will be on a big BBS-2. I have Winkie experience, and the French guy has drilled BBS-2. Of course, you know who goes where. It's just like the army – the sharpshooter plays in the regimental band and the tuba player is on the firing range hitting everything but the target.

The young fellow struggles with the Winkie but soon catches on. The old guy struggles with the BBS-2 and catches on much more slowly. I am drilling "A" holes with a bullnose bit (no core) and I am out of my depth, figuratively, not literally. The holes are not deep, only 100 to 120 feet, but I am not used to either the water pressure required by the bullnose, or the power produced by the six-cylinder diesel. I immediately burn a bit. (Lack of sufficient water causes the bit to heat up and burn off the diamonds.) At least I am smart enough to back off before I weld the bit into the rock, as that would have caused more problems. Part of the string would have been lost and the hole would have to be abandoned. I go to get a new bit and the foreman is very disappointed, to say the least. I get hollered at. Drillers must be in short supply or I would have been fired on the spot.

I start to get a handle on things and although my production is iffy, the company keeps me on. We work ten twelve-hour days with four days off, and the money is pretty good. On the evening of the tenth day I head for North Bay. I will be getting married tomorrow.

We have already bought the marriage licence and have made an appointment with a Justice of the Peace in Parry Sound, a town on Georgian Bay an hour south of Sudbury. (Bobby Orr is from Parry Sound, and is already changing the way hockey is played in the NHL and the world.) The reason we are going there is that it is the closest place we can find a Justice of the Peace to marry us. As unbelievable as it may sound in this day and age, at that time few J.P.s were willing to perform a "mixed marriage." (She is Catholic, I am not.)

So we take an early morning drive to Parry Sound, walk into the town office and plunk our license down. The lady takes a look at it and says that it has been filled out wrong and we can't get married today. What? Now what do we do?

We have to buy a new license, she says. Then it goes to the registrar in Toronto for approval and that can take up to a week. It's no use pleading our case – we will just have to wait. What the heck – we may as well have lunch and return to North Bay.

We go to a semi-fancy restaurant for lunch. The waitress sees we are a bit down in the mouth and asks us why? We tell her our woeful tale.

"No problem," she says, "I know just the lady to talk to."

A quick phone call to her cousin in Toronto results in the waiting period being waived and Marrying Sam performs the nuptials, complete with flowered waistcoat, watch fob and little-old-lady witness with tears of happiness in her eyes. We go back to the semi-fancy restaurant for our wedding supper, but our waitress/benefactress has gone home – so we leave her an envelope with a little gift for service above and beyond. We have a three-day honeymoon and I go back to Mountain Chute early Tuesday morning.

It's now November first. I go back to the job for three more fourteen-day swings and it's just too dangerous – even for me. I am drilling on the side of a bare outcrop hill, and the drill is moved from one precarious set-up to another. We have to build log cribs anchored by rock bolts, and in sub-zero temperatures, the return water freezes as it runs down the hillside. On December 12, I take a fifty-foot slide down the rock face, trot over to the foreman's office and turn in my time. My drilling career is over for now

As I leave the Ottawa Valley I make another of my famous bad transportation decisions. The Olds is getting tired, and I'm now a married man, so on the way through Pembroke, I trade the Olds in on a two-year-old Volkswagen squareback sedan. Married men should drive an economy car - right? Wrong – four months later the Volks would suffer a major heart attack – but that's another story.

For now, at least, I'm satisfied. The squareback is upscale compared to the Bug, with an attractive interior, adequate legroom and with an auxiliary gas heater, it is nice and warm inside. I am used to keeping a pack of smokes on the broad expanse of the Olds dash, and the first thing I do on the road back to North Bay is reach for a smoke – and bang - my knuckles hit the windshield. My world is shrinking, it seems.

I get back to our little apartment, and my wife is surprised, but glad to see me. She understands why I quit and even though she is working, I need a job as well, so I put my pride away in a safe place and call Inco.

Herb hires me on the spot. I will join my old friend Barry at Smooth Rock Falls on Jan 2nd, 1966.

Back to the bush again!

Table Scrap: Car pool (I should stick to the shallow end.)
The drive to the Mountain Chute project is about 25 miles through rolling farm fields and maple woodlots. It is not onerous and in summer, and especially in fall it would be downright beautiful. Five miles from my bed-and-board I pass through Mount St. Patrick, a warm, old-timey village – very pleasant. Mr. O'Reilly tells me there are a few guys from the Mount at the Mountain Chute project. He says they car pool – perhaps I can join them. I know one of the guys, and at work the next day I broach the subject.

Sure I can. He is already up to speed, thanks to Mr. O'Reilly.

(So, I've built the backdrop – let's set the stage.)

This is predominantly Irish country. Heck – It's predominantly totally Irish. I am supposed to have Irish roots, but these guys have their papers and ear tattoos.

Secondly – I am built sort of mix-and-match. My brain has standard straight-cut gears – no synchromesh. My mouth, however, is fully automatic and slips into gear far too easily.

So I join the car pool at the Mount, and introductions are in order. One guy's name is Kenneally,

"Glad to meet you," says I, "You must be Italian."

There are a few snorts from the others and Mr. Kenneally transforms into Black Irish! The feng shui in the car is not good.

For the rest of the seven-day swing he sits in the right front, I sit in the left rear.

It's his turn to drive next week, And Mr. Cannoli does NOT invite me into the pool.

Sidebar: Carl will spent the winter and summer of '66 on his Winkie drill. In October he does a couple of holes for Inco near Thunder Bay and is in quartzite for almost every foot. He is sick and tired of breaking even, and Inco's own Winkie drill program is in trouble. He sells his drill and signs on to Inco for five years to run their Winkie department. Good for him. I'll see Carl often over the next few years.

Carl's winter drill camp - pretty basic.

Winkie lake-ice setup

Sidebar: In the '60s Carl lost his temper and fired a guy. A couple of months later Carl is in a Sudbury bar and the guy spots him and heads for his table. Carl sort of expects a punch in the nose, but the guy sits down to talk. "I don't really blame you for tying the can to my tail," says the guy, "I wasn't doing the job, but I was trying. You have to understand, that some people catch on a little slower than others."

Carl said that he felt like crap, and still does when he thinks about it. A few years after that Carl started up a hardware business. Twenty successful years later he sold the store, and his first hire was still working for him.

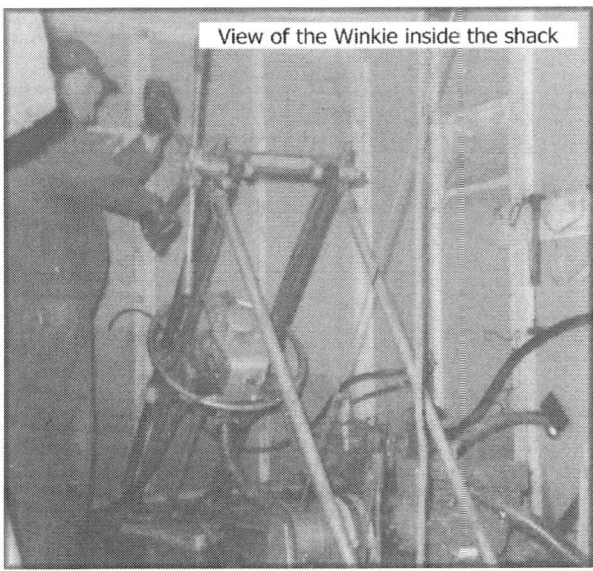
View of the Winkie inside the shack

Chapter II   1966, 67, 68

Smooth Rock Falls - Camp 20
On January one, 1966, me, my packsack and eiderdown are heading to camp 20, south of Smooth Rock Falls. Camp 20 is an old Abitibi Paper bush camp and Barry has used it for a couple of seasons. Back in '62 I had worked for Barry near Cochrane, if you recall. Over the last two years the work has moved westward, and this year the area will probably be wrapped up.

It's a comfortable camp. A long kitchen/bunkhouse is flanked by a number of smaller buildings. Barry's office/living quarters is in one, another houses the helicopter pilots and mechanic, and another is for the geologists. The core shack, and other equipment storage/junk drawers are on skid row.

The main building is where I will bunk. The west end holds the kitchen and a huge table. A door from the dining room leads to our living quarters, basically a dormitory with two rows of beds and a drafting table at one end. There are 18 to 20 men bunked in here, three or four anomaly chasing crews and other support personnel. There are no showers, no partitions and no privacy at all. I miss the camaraderie of a smaller camp, but the food is good, the beds are comfortable, and a job is a job.

There are a few familiar faces, some I have met before and others I had heard of. I do meet two men I like right off the bat. One is Brian, who is barely out of his teens and supporting a widowed mother and younger siblings back in Timmins, He is a thoroughly remarkable young man and a few years ago I had the pleasure of joining him for a barbecue on Don MacEachern's patio. Brian and I were never close, but we sure enjoyed that afternoon of swapping lies.

Sidebar: Don MacEachern is another story waiting to be written. After his Inco career, Don went on his own doing contract staking and geophysics. He married a Fort Frances girl, raised a fine family, and over the years remained faithful to his chosen profession, through thick and thin times. Don and a partner were the initial stakers of our soon-to-be Rainy River Valley gold mine and helped nurse it along to its present 6.5 million ounce resource estimate. Sadly, Don passed away before this book was published. But we sure enjoyed sharing old lies on his back deck.

Besides us anomaly chasers, there are three big drills working out of camp – thus, we have two choppers. They are Bell 47 G-4's and have a lot of lifting power compared to the G-2 and G-2A. Lack of roads and lakes in the area means that all drill moves are done by chopper, so they are kept busy. Our pilot is ex-U.S. Air Force, and he doesn't like me very much. That's OK, I don't like him either. We both do our job and stay out of each other's hair.

Don and I share a chopper for our daily in-and-out. It's the same deal as I have previously described, two men into each anomaly, followed by the one man/one man routine, with the afternoon trip in reverse.

One day we are sitting at our tea fire when we hear something coming toward us through the bush. We think it could be a moose, and it is – a moose named Don! He has had EM equipment trouble and has left his crew to do some mag work while he takes a four-mile walk to have tea with us. He helps us on our grid that afternoon, and at 3 pm he says he will head back, the chopper will be coming soon. I point out that the chopper has to make three trips anyway – why not wait here? Don says he needs the exercise and disappears. I scratch my head, but I will learn that Don is Don. He is like Pass-Me-da-Hax Henry, without Henry's wild streak. An eight-mile hike through trackless bush is nothing to Don, and his work ethic is exceeded only by his stubbornness and honesty – a man to be admired.

The work routine is old-hat Inco – work 13 days and take one day off - one day per week is banked for accumulated time off at season's end. That leaves one day left over, and Inco says you work that day for your room and board. Remember 1960 when I was told I would make all that money with free room and board? Well – this proves that nothing is free in this life - free is just an overused adjective.

You may also remember Harry and Norm – the two recently married men unhappily separated from their wives. I have now joined the club. I do have it a little better than those other guys, though. I get to see my wife for a few hours every second week. On the last day of the 13-day stretch I have a quick supper and hop in the Volksmobile for an eight-hour drive to North Bay, arriving after midnight. Twenty-four hours at home and I head back to camp at midnight, and as I pull in to camp at 8 am the chopper is warming up.

It could be worse. Barry is a decent guy, and while our weather is good, we do have the occasional snow day. There is always something to be done around camp and Barry never calls a snow day a day off so our fourteenth day comes up as regular as clockwork.

The auxiliary heater on the Volks has been acting up, and on the road, without the heater running, cold air will flow through the heater into the car and I have put a piece of cardboard over the inlet to prevent this. On the thirteenth day it is my habit to start the car and let it warm up while I grab a bite to eat before heading south.

This time I jump in, fire up the Volks and head to the bunkhouse to change clothes and eat. Without thinking, I had turned on the gas heater. Nine times out of ten it will misfire and time out. I guess this is the tenth time, because as I'm filling my face someone comes in and tells me my car is on fire. I run outside, and smoke and flames are pouring out from under the hood. I whip the hood open, and it's easy to see what has happened. The piece of cardboard has blocked off the air from the heater. It is red hot, has melted the plastic line from the fuel pump to the heater, and with each pulse of the fuel pump fresh gasoline is sprayed on the fire. I shut off the key, throw on a few shovelfuls of snow, and disaster is averted. Everyone thinks it's pretty funny – they all laugh like crazy.

There is no real harm done other than some paint discoloring. I cut off the burnt end of the fuel line, reattach it to the heater and pull off the ignition wire just in case I have another brain cramp. A few weeks later I will regret dousing the fire. I would have been money ahead had I just let that little Teutonic monster burn to the ground.

In early April I have had it, and so has my wife. After one day off, I simply don't go back. I drop a note to Barry and ask him to ship me my eiderdown. Barry understands – he is married himself.

My wife gives her notice, and two weeks later we head for the farm near Fort Frances with all our worldly possessions in the back of the Volks. A two and a half year hiatus from the bush life lies ahead.

How to Paper-train a Bushwhacker

The trip along the North Shore of Superior is great. It is sunny, not too cold, and the highway is bare. We pass Fort William/Port Arthur, and by the time we reach Atikokan it's obvious that one heck of a snowstorm has recently passed through. The pavement is snow-packed, and the closer we get to Fort Frances, the higher the snowbanks are along the road. It turns out that five days ago a Colorado low had veered up along the north-east side of Superior and dumped five feet of snow on the Rainy River Valley. My dad was born in the 1890's, and he said it was the first time in his life that he was been snowed in. (Of course, they always had horses back in the day. Dad still has Dolly, his last colt, but she is now 25 years old, living a well-earned life of luxury on her timothy pension.)

After the storm blew over my brother had delivered some groceries to the farm, carrying them in the three-quarters of a mile side road on cross-country skis. By the time we arrive, the municipality has hired bulldozers to open the road and things are more or less back to normal.

We don't arrive by Volksmobile, though. Twenty-five miles east of Fort Frances it turns a main bearing and dies, proving the old adage is right; if it has tracks, tires or tits, sooner or later it will give you trouble. The Volks arrives on the hook, and we arrive courtesy of my brother.

First things first. The local paper mill has gone into seven-day production and I head down to the personnel office and hire on. The personnel manager tells me that every time he hires a country boy he gets static from the town boys.

"Then why did you hire me?" I ask.

"Well," he says, "If I didn't hire you country boys, we'd never get any work done around here."

I buy a beater for basic transportation and we rent a little house in town. I go to work on a seven-day swing shift and help out on the farm on my days off. In May our first daughter arrives and she is a sight to behold, all ears and a great big smile. As I hold her for the first time, I look at this perfect little girl and wonder if I am up to the task ahead. (I don't know if I had that much to do with it, most likely it was good genes and a good mother, but she is now a perfect big girl.) In January of '68 our second daughter arrives, another perfect little bundle of joy. Our little family is complete, and that spring we buy a fixer-upper twelve minutes out in the country, only four miles from the farm, and things are pretty good. My wife is working at the hospital in town and her mother moves in with us. We now have a live-in babysitter, my wife has a semi-reliable car to drive to work, and I have an old ¾ ton Dodge long-box to get me to the mill on time. I'm still only half-tamed, though. I try to be a good husband and stay-at-home father – I really do, but cracks are starting to appear in my facade.

The job is part of the problem. The pay is decent enough, and we are both working, but there are limited opportunities for personal initiative in the mill. Punch in, punch out – that's about it. I like working in the wood room end of the mill – it's clean and there is always the smell of fresh river water and fresh wood. Sometimes I am put on the paper machine end.

Over there, it is smelly and noisy working with the boys from town. The paper machines are smelly and noisy also. I like to get a full paycheck, but I tell the schedule supervisor to leave me at home if the machine end is the only job available that day. He is a pretty good guy and I don't miss many shifts.

Another problem is that the paper company has a lock on the local labour pool. They are the best paying job in town, and because of that, there is always someone willing to take your job if you don't want it, and the mill can tend to be a bit arrogant. They run a "spare board" of at least 120 men and after two years I still can't count on a steady shift. Us spare guys get to cover holiday time mostly, and naturally, the old-timers take their holidays on their midnight swing, so guess who gets to work mostly midnights? Us spare-boarders, that's who.

I buy into the union concept wholeheartedly in an effort to make the job more interesting. I feel that the union is necessary to protect workers' rights and I attend every meeting I can, but some guys drive me nuts. Some old guy who has dead-headed on a soft cushy job for five years now, wants another soft cushy job, and he can't understand why a guy with less whiskers is ahead of him in the line of progression. The argument goes on for the whole meeting.

In an attempt to break the midnights-only mold, I bid on a slasher bulletin and get it. The slasher is at the head end of the wood room. Eight foot pulpwood is fed to the slasher table by the yard chain or river chain. There are three of us on the slasher. As the wood comes over the spike roller it is our job to keep the 8-footers lined up on the feeder chains, which carry the logs ten feet or so to be slashed into four foot lengths. From there, the wood is fed into four huge drum barkers and after the bark is removed, the four-foot logs go up to the charging floor on a rubber conveyor belt. On the charging floor they are fed into four foot square pockets above the grinding stones, then the ground up fibers are piped over in a slurry to the paper machine building to start the metamorphosis known as "making paper."

The slasher is hard work and can be dangerous, but I like it. I last three months and hurt my back. I am out of shape compared to my bush days, and although I take a couple of weeks off, I have to face it – the slasher and I are not compatible.

A few weeks later a slasher guy gets sick and has to go home. I am on the sorting belt, and the foreman, an old-timer, tells me he wants me to fill in on the slasher table. I tell him I had to drop the job because of my back, and he says, "If you were any kind of a man you'd go up there."

"Bill," says I, "It's because of assholes like you that we have a union."

He doesn't like it much, but tough titty, Bill has to call a man in on overtime.

Table Scrap: Keeping your Wood Box Full on Company Time.

I worked with all the slasher guys, moving from crew to crew covering holiday time. They were a town-and-country mix – all good men. The job required strength and agility, and one of the strongest and most agile was Mervyn Caul.

A lot of folks burned firewood in the '60s. No outside stoves yet – perhaps a supplementary stove in the basement, and for sure one in the cabin or garage. I had a coal furnace at home, and the occasional addition of wood helped out with the coal budget.

The pulpwood came into the wood room in 100 inch lengths. The slasher saws divided each stick into 48 inch lengths, leaving small end blocks. These dropped into a chute leading outside where they would be periodically loaded into a company tandem to be delivered to the landfill site north of town.

So we would park our pickups beneath the chute and take our wood home after the end of the shift. A lot of us did this, not just the slasher guys – one or two foremen collected trim blocks also. It was not a scam, we actually saved the company some truck and loader time.

One night my Dodge was under the chute and I was slashing with Mervyn and another guy. We were taking on poplar from the yard chain and it was coming in at a slow pace. Poplar is hard to peel and has to be recycled through the drum barkers so only one flatcar (40 cords) was being unloaded. This would take up to an hour or so before we switched to spruce and jackpine. Whoever had shipped the carload had sneaked in quite a bit of birch with the poplar.

"Do you want some good firewood?" Mervyn asked me.

So every time a birch log came in Merv would keep pulling it back to the trim saw until I had a load of stove-length birch in the dodge. It was a dangerous thing for him to do but the foreman never caught us.

It's nice to readjust to a more or less normal life. At the mill I work with old friends, make new friends and reconnect with the Rainy River Valley. I get to hear new common room (coffee table) stories and get to make up some of my own.

Table Scrap: Terry and the Dam Ladder

Terry Robinson – I've known him since 1950. Farm raised and farm forged, he was a good, strong, hard man. Or, as we used to say, good for nothing, strong on smell and hard on grub.

Terry went into the mill after high school. One of his first jobs was at the Crilly power dam on the Seine River east of Fort Frances. I don't recall exactly what they were doing – probably maintenance.

The job took longer than anticipated. It was late October, winter was just around the corner and freezing rain had covered the top of the dam with ice. Terry was on a ladder hanging on the dam face. He was scaling spalled cement and ice.

The ladder was suspended by a rope which extended over the top of the dam, attached to a point on the far side. Another man was chipping ice at the top and he cut Terry's rope!

There was no point in bailing out – he was sure to suffer injuries on the rubble below. Terry rode the ladder down. Lucky for him it was a wooden ladder. When it hit bottom Terry broke the rungs – one at a time. Each time a rung broke he slowed down arriving at the bottom with only two speed reducers left. He gave one sore knee a rub and went back to work. I assume everyone laughed like crazy.

Table Scrap: Terry and the Teacher

Terry didn't like high school much. Who did? A lot of us treated it a s a place to hone our social skills rather than actually learning something.

One day a teacher cornered Terry. "Why don't you buckle down and study once in a while? He said, "Otherwise you'll end up driving truck like your brother."

Terry's brother doesn't drive much anymore. Someone has to keep track of the fleet, the loaders, the crushers and screeners, and more than one large gravel pit.

But we are just too darn close to the Shield. There is exploration activity all around us, and I can't help running into people from my past life. My wife is patient and understanding, but I am like Oscar, back in '62. She can see me flying a short distance now and then, and is probably wondering when 'm going to join the migration.

I get a phone call one evening. Gerry (Mining Corp) is in town. He has Raymond and another guy with him and they have a couple of grids to do in the area. I join them at the bar and we have a great time. They ask me about this bug that they have never seen before, and they describe it to me. "Why, that's just a wood tick," I say."

They want to know what a wood tick is. They've never seen one.

"No big deal," I tell them. "It latches on to you, fills itself with your blood until it is the size of a glossette raisin, lays its eggs and drops off." Of course I have to throw in a little more BS. I tell them the old wives' tale of how you have to make sure you don't leave the head in your skin when you pull the tick off, otherwise the head will keep burrowing into your body until it reaches your brain. Then I laugh like crazy at the look on their faces and tell them I am joking.

The next day after work I stop at their hotel and find that they have checked out and left town.

(A couple of years later I run into Carl, my boss back on Nighthawk Lake. He tells me I scared the bejeezuz out of them. Gerry called Toronto and told them the whole Rainy River Valley was flooded by beaver dams. He and the crew boogied on eastward, and never returned to wood tick country.)

In September '68 I run into Jack Martin in town. Jack was my party leader in Chibougamau and although we were never bosom buddies, we always got along well with each other. We are very glad to meet again, and sit down to catch up on the last six years. Jack is now working for Noranda Explorations out of their Port Arthur office and is Noranda's geophysical foreman for North-Western Ontario. He is on a small job just off the east side of Rainy Lake and is camped with a helper and line cutter on Red Gut Bay. On this day they are staying at a hotel for one night and will head out to the bush early tomorrow. I have a great idea – why not come out tomorrow afternoon? I will meet them and they can come out to the house for a barbecue supper. Jack says he will supply the steaks.

After work the next day I go to the bar and find Jack sitting with Sam Duggan, who owns the drilling outfit currently on contract to Noranda. Jack doesn't look very chipper, and they are not talking much, just sitting there looking at their drinks. I sit down and keep my mouth shut. Obviously something bad has happened and over the next hour the story comes out.

Jack and his two men had headed out to camp very early that morning. They had a 16' aluminum boat with a 25-horse motor and were cruising down Red Gut with the helper at the tiller. A gust of wind blew the helper's hat off, and without thinking he cut the motor rather than throttling down gradually. The backwash came over the transom and swamped the boat, sinking it stern first. The three men were in the lake and because they were using their life jackets as seat cushions, the jackets floated away. The boat turned over and the built-in styrofoam under-seat flotation prevented it from completely sinking, but with the heavy motor on the stern, only the upside-down bow remained above the water.

They hang on to the keel rib and consider their situation. The water is pretty darn cold in September and they are dressed for work, with heavy bush boots and jackets. Shore is only 300 feet away, and the helper wants to swim for it. Jack says stay with the boat – someone will come by pretty soon. The helper won't listen to Jack, heads for shore, makes it halfway and disappears beneath the water.

Jack and the line cutter hang on, but the cottagers have gone back to town and no boats go by. Hypothermia is setting in, and the line cutter starts to drift in and out of consciousness. He loses his grip on the keel a couple of times, Jack catches him and keeps talking to the guy, telling him to hold on, trying to keep him awake. But Jack is barely able to hold on himself, and finally the line cutter slips away. Jack hangs on for another hour before a passing fishing party rescues him. Jack tells me it is the longest hour he will ever endure.

"Closh-closh" says Jack, "All I hear is the sound of the ripples hitting the boat. Closh-closh, closh-closh."

I return home to sit quietly with my family and consider the tenuous hold we have on life. One day here – the next day gone. Crap happens, that's for sure.

Table Scrap

I will recall that day many times in the next few years. One instance in particular stands out in my memory. I was pulling a three-man camp out of a smallish lake east of Red Lake and had gone in with the Beaver to show the pilot where the camp was located. I thought we would do two trips, but we keep stuffing gear in the Beaver and the three men climb in on top of the load, their heads brushing the fuselage ceiling. I'm riding shotgun. (There are some perks when you weigh 250 lbs – I get to sit in front.) The lake is narrow, with high ridges on both sides. It's long enough for the Beaver, but we are heavily loaded and we have no headwind. The east shore is a low valley and the pilot taxis down into a creek at the west end to get a maximum takeoff run. We hang up on a rock. I jump into the water, lift the float a little, point the plane in the right direction and climb back in. I have a lot of confidence in this pilot's ability.

We pull out of that lake with room to spare, the Beaver banks left in a U-turn towards Red Lake, and at 400 feet we run out of gas!

There is nothing in the world quieter than a dead-engined Beaver, and I can hear it – Closh-closh, closh-closh. The hair on my arms stands on end, but for some strange reason I am not all that concerned, probably because the pilot is so calm about the whole thing. He just reaches down, selects a full tank, gives the primer handle a couple of pumps, the engine fires up and we are on our way.

I never did rat the guy out – what would have been the point? But what if that tank had run out two or three minutes earlier? Closh-closh, for sure. One of the guys in the back didn't share my confidence in the pilot – he threw up in his boot.

Sidebar: Boots are the bush-whacker's barf bags. Hang around any float plane base and sooner or later you'll see some guy hop out of the plane on one stockinged foot and wash out his boot in the lake. Then you laugh like crazy – it's the law.

ll is not well at the mill. The company has been messing with the union contract, and on October first we shut her down, or at least my union does. The rest of the mill

keeps working. We can't really blame them, it's an illegal strike (wobble) and they have mouths to feed.

We set up picket lines, but we know we can't enforce them. Blocking access to the mill will only result in some time in the slammer, so we don't really have picket lines, just information lines. The only union that won't cross our lines are the Teamsters from International Falls, our sister town and paper mill partner just across the river in Minnesota. The Teamsters run the trains daily across the International Bridge to pick up and deliver the rail cars – and Teamsters never cross picket lines – never, ever!

The next day we discuss this at our rented strike headquarters and agree to give the Teamsters a free pass. They might as well cross our picket lines – everyone else does.

My brother is in a bit of a quandary. He works in the mill office, and salaried personnel are assigned to run our end of the mill. He tells his supervisor he won't do it – I am on the picket line and my brother doesn't want to cross. His supervisor understands, but tells him he can't cover for him for too long. My brother is upset, naturally, and I say I'll talk to the boys at strike headquarters.

The next day I bring my brother up. Will they guarantee him a union job if he is fired? All I get is blank looks, so I call my brother and tell him he may as well go ahead and cross. These assholes won't lift a finger to help themselves, or him.

I'm starting to rethink the whole union scenario, but I don't have time to worry about it right now. After 2½ years of punching time cards I'm having fun again. I'm a picket line captain, which means I just have to make sure I have enough sign holders to cover my four hours of designated picket duty.

The local TV cable supplier is restringing new cable in town and me and my trusty ¾ ton are pulling reels of coaxial cable up and down the streets. Sometimes I even get to strap on climbing spikes and drill holes in hydro poles (always lean out, always – otherwise you will slide down the pole and spend the next few hours pulling splinters out of your belly button.)

And – there are still a few schools and businesses in town burning coal, and the local coal delivery guy has retired. When we are not TV cabling, me and Mr. Dodge are unloading boxcars of coal and shoveling it into coal chutes around town. I am working sixteen hours a day and enjoying every minute of it.

Table Scrap: Pulling Wrenches, Fighting a Forest Fire and Ducking a Grand Theft Canoe Rap – All in One Day.

The weather was great for the first three weeks that October – an extended Indian Summer, and although I was busy-busy my weekends were free.

Bobby Green had a nice 1958 Meteor, but the motor was a bit lazy, He needed a valve job. Money was tight, so we decided to fix it in my garage.

We pulled the head off the six-banger and dropped it off at West End Motors. We would pick it up after West End did the serious stuff and bolt the Meteor back together on Saturday.

Now we have to go back two or three months to June or July. My wife was working midnights and I was at home sleeping. Two of my Yahoo buddies were indulging at the Emperor Hotel bar and left at closing time. They walked past a pickup in the parking lot, saw a canoe tied on the pipe rack and took it with them – just good ole boys with brain cramps.

They drove around for a couple of hours, realized they had a canoe which they didn't want and they decided I needed it. My wife came home in the morning – I was having my first coffee in the kitchen. She asked me where I got the canoe. "What canoe?!"

Sure enough, I had a cedar strip in my front yard. Shortly thereafter one of the shoplifters called. After hearing the story I suggested they should get the darn thing out of my yard. They told me not to sweat it – there was no way the canoe could be traced and when I returned from work that day it was gone. The theft was reported in the weekly paper but I was unworried – I was off the hook. (???)

So on Saturday morning in October Bobby and I put the head back on his motor. We had it torqued down and had just bolted on the valve train when my mother-in-law bolted into the garage. "The field is on fire!" she yelped.

Our house stood alone – one quarter mile east of the village and another quarter mile to the next neighbour east of us. Our yard was surrounded by tall dead grass, dry as tinder and now burning vigorously. My woodpile was in danger and the blaze, driven east by a brisk wind, was picking up speed.

I told mom-in-law to call the Lands and Forests and Bobby and I grabbed two pails of water, a couple of gunny sacks and shovels. We saved the woodpile and were about to head east when the L&F crew arrived. They had six men, a Wajax and pack pumps. We were told to relax and leave the rest to the pros.

By now it was 2 pm. We were greasy, sooty, sweaty and coming down from an adrenaline high, but we figured we'd set the valve lash and button up the Meteor before taking a break. By 3 pm we were finished.

We deserved a cold one before washing up and Bobby likes a glass, which he got. He sat at one end of the kitchen table, I sat on one side and my two giggling little girls sat opposite me nursing their kool-ades. Our tomcat entered stage left.

He was a big old gentle guy. My girls carried him around draped over their arms – front feet and head hanging down one side, back feet and tail dragging on the floor.

The cat sat and stared at Bobby. "That cat does not like me!" said he.

I scoffed at the thought. "That's a real pussycat," said I. "He wouldn't hurt a flea."

Just then the cat launched himself at Bobby's throat. Bobby hollered and went over backward throwing out his arms and I got the glass of beer in my face and on my tee shirt. The cat headed for low ground in the basement, I wiped some beer off my face with a greasy hand, we all laughed like crazy, and there was someone knocking on the door.

I answered the door, and there stood a cop. "What do you know about this canoe?" he asked.

"What canoe?" said I. It was deja vu all over again.

"The L&F crew found it. It was reported stolen 3 months ago and the serial numbers match."

Well holy cow – the Yahoos had just pulled it into the scrub between our place and the neighbour's.

The cop accepted my plea of non-involvement. I guess he figured that anyone who looked and smelled that guilty had to be innocent.

And that's how you do a valve job, fight a forest fire and snaffle a "Get out of jail free" card all in the same day.

Dang it all, the mill lays down the hammer and here's the deal. The company plans to build a kraft mill and if we are good boys and go back to work they will build it here, otherwise it will be built in Kenora, a town on the Trans-Canada, 110 miles north of us, where the company has another paper mill. We have a full turnout for the vote and it's almost unanimous. We will burn our placards and tuck our tails between our legs. Of course, I am part of a small minority, and not a very vocal minority at that. Majority rules, and on November 1st I am back on the time clock.

It's going to be tough, I can tell. Everybody knows my stance on the issue, and I'm pretty unpopular in the scheduling office. I can see many months of midnight shifts ahead. I have heard that Jack, who understandably is having a difficult time dealing with the tragic events of September, has packed it in and Noranda is looking for a new geophysical supervisor. I talk it ever with my ever-patient wife and she agrees. I call Noranda's area geologist in Thunder Bay, and two days later, after he has made a few calls himself, I get an answer: I am hired on the spot. In mid-November I bus to Thunder Bay, meet Don Cross, and drive home in my company truck. It's back to the bush!

Fun and Games in Bridges Township
I dig out my packsack, moccasins and bedroll, give my wife and girls a smooch and head for the job. Noranda has a hundred claims in Bridges Township, on the Trans-Canada, 17 miles west of Vermilion Bay. We will be staying in two single room cabins at a nearby restaurant/gas bar with two beds per cabin, and will eat our meals in the restaurant. I meet Nick, the line cutter, and tomorrow one man will arrive by bus from Port Arthur. another man, Mel, is already on his way with his own vehicle.

The grid is along the north side of the highway and is 3 miles east/west, and up to two miles wide. The west end is right across the road from us, and most of the grid is easily reached by road. The line cutter is almost finished and one of my men will help him with the last of his chaining.

The survey will be no problem. Noranda uses the JEM and McPhar fluxgate mag, two units I am familiar with. When Kenny arrives from Port Arthur I find he is experienced on both, so he will be my mag man.

Kenny and I set out to read base stations. The first two miles of base line goes well, but as we extend our readings on to the last mile, something seems amiss. We do some backtracking and eyeballing. Then we put up some line pickets, and sure enough - the easternmost mile of the baseline takes a 10 degree turn to the south!

Nick, the line cutter, is quite a character. When I tell him his base line is crooked, he blames it on that darned raven! This is a new one on me. Nick says that a raven landed on one of his baseline pickets and moved it. When he realized his base line had a bend in it, he had already cut half the lines, so he just continued on. The north lines from that point are now in a 10 degree fan, and the gap gets pretty wide at the north boundary.

Oh, well - I'm not about to make Nick recut and chain a third of his contract. We will fill in with pacing and surveyor's tape. If the instruments come up with anything interesting, we can cut and chain later. I lodge a formal complaint with Mother Nature anyway, but the raven never returns my call. I guess he has other baselines to adjust.

Mel will not be here long. He has been working with a company prospector all summer and is due to take the winter off. He is only here to help Nick finish his chaining, then he will stake some claims to fill in an open corner and leave for home. I tell Port Arthur I need another man, and three days later Ahmed the Terrible gets off the bus. You may think I am too hard on Ahmed, calling him "The Terrible," but let me tell you, he is the screwiest guy it has ever been my displeasure to run into. I'll tell you about Ahmed and you can be the judge.

He looks pretty normal at first, not a very big man, and pleasant to talk to. He is an immigrant from the Middle East and tells me he has a degree in geology from a university (name withheld) in (city name withheld) in (country of origin withheld.) Sorry for all the brackets, but he may be kicking around somewhere. I may have been born at night, but I wasn't born last night.

Kenny is on the mag, so Ahmed will be my JEM helper. I show him the unit and he nods his head. "I know, I know," says Ahmed. I start to tell him how we operate the unit, and again with the "I know, I know." This little guy is very agreeable, I think, and we head for the picket line.

The JEM operates on a "shoot-back" method. The two operators are separated by 200 feet and the resultant reading is at the midway point, so I plant Ahmed at the baseline, tell him I am going to 2 north, and he is to wait for my call. I walk 200 feet and stop, and damned if Ahmed doesn't run into me! "What the hell ae you doing here?" I yelp.

"I no stay back there," says Ahmed. "Maybe wolf!"

Lord love a duck! This guy is afraid of the bush! I consider where he is coming from, and I guess this is his first time away from the shifting sands. I light a cigarette and tell Ahmed that we are within a few hundred feet of a trans-continental highway, and to date neither the line-cutter nor any of us has seen a sign or a track in the snow of a predator.

I start down the picket line again, but this time I'm watching. Ahmed starts to follow. I turn around and point a finger at him. "Stay!" He won't stay. I turn around and walk him back to his spot again. I point to the picket.

"Ahmed," I say, "Let's pretend this is a steel post standing in cement. There is a chain attached to the post, and the other end is attached to a ring around your ankle. Now let's pretend that if you move one foot from this post, I will come back here, pick you up and toss you in the bush!"

Ahmed looks at me and says, "You no my boss. I your boss. I am geologist!" Well, now I've heard everything, I think. (Not true, though. I have more to learn. Ahmed keeps a notebook in his packsack, and one day I snoop. Inside the front cover is the sum total of his geological knowledge. Under the heading of "Granite" are three sub-categories: "Pinky, Whitey and Grey." Geologist my ass!)

It's 3 pm and I've had enough for one day. On the way back to camp we catch Ken at the end of a mag line and I tell him to come with me to town for a beer. I need some R&R (respite and re-evaluation.)

In town, I find a pay phone and call Port Arthur. There is a phone in our restaurant, but it's a party line, and besides, Ahmed is ever watchful and nosey.

I tell Don about Ahmed the dipshit, and he says "Work with him. It'll turn out OK."

I have a flashback to '62 – where have I heard this before? Well, maybe Don is right and I need to be patient and understanding. After all, things didn't work out that badly with Dean, the hearing impaired guy.

Before we head back I stop at the local Co-op store and buy two boxes of Christmas oranges, one for Ahmed and me in our cabin, and one for Ken and Mel. The restaurant food is very good, but fresh fruit is always nice to have around.

Back at the cabin I show Ahmed the oranges and tell him to help himself, no need to ask – just take one anytime you feel like a snack. Ahmed looks a bit perplexed. Maybe he has never seen an orange before. Yeah, right – I chuckle at the thought.

That night I am awakened by one hell of a commotion. I sit up, turn on the light and Ahmed is standing on his bed, hammering on the wall and speaking in tongues. I tell him to go to sleep for Christ's sake, and he settles down. I decide to never turn my back on him, and I sleep very lightly after that episode.

The next morning, as is my usual habit, I have an early breakfast. Ahmed comes in a little later looking hung over, and this is strange – Ahmed doesn't drink. I go back to the cabin to get into my bush clothes and there is the orange box – empty! Has Ahmed the squirrely been stashing oranges? I check under his bed and pillows – no oranges. I move the wastebasket to look into the little closet, and holy hell! The waste basket is full of orange peels. The little freak ate the whole box! No wonder he was climbing the walls last night! He had OD'd on citrus! Well, things could be worse, I guess. I guess wrong.

We give the JEM one more shot, but it's hopeless. When I do get Ahmed rooted to his spot, he has no concept of orientation or reading – no concept at all. At noon I give up and tell Ahmed we are taking the afternoon off. We return to camp and have lunch. That's one nice thing about this grid. We are so close to home we never have to pack a lunch, just walk back to the restaurant.

Our cabin is pretty basic stuff, just two single beds, a map table and a small wash stand beside the door. On the wash stand there is a tin basin and a pail of water, something to dash on your face in the morning before breakfast. There are public washrooms in the gas bar, with a sink and hot water for washing and shaving. We have permission to use the owners' private facilities if we want a shower, but most of us use it sparingly – we feel we are encroaching on their privacy.

But Ahmed likes to have a daily shower. This wouldn't be that big a deal if Ahmed was a normal person, but Ahmed's choice of attire when in camp is pyjamas – always pyjamas. Every evening he trots through the restaurant and kitchen in his pyjamas and slippers, and Bill and May, the owners, raise their eyebrows. I think they don't like it much. Neither do I, and I talk to Ahmed about it, and get my usual blank stare. I apologize to the owners and they say it's OK, it's off season, and we are business they would not normally get this time of year.

So today we are taking the afternoon off, and Ahmed wants to take the company truck to Dryden. Someone has told Ahmed that the Viking Motel has a pool and sauna, and Ahmed wants a steam bath. I figure he must already be the cleanest little Arab in the Western Hemisphere, but I have a brain cramp and give him the keys. I go up to the restaurant for a leisurely cup of coffee, and 45 minutes later Ahmed comes in. I figure he must be getting ready to go, but Ahmed sits down and says, "I no can find."

"What do you mean, you no can find?"

Ahmed shrugs and says, "I don't know, I drive and I no see."

"What do you mean you no see? The Viking Motel is two miles west of Dryden, right on the Trans-Canada. You have to see."

Ahmed says. "I no see. I drive to Dryden, I drive back, I no see."

I almost fall off my chair. I check my watch and I maybe have sat at my coffee for an hour at most. It's 42 miles to Dryden, so the little freak has made the round trip of 84 miles in an hour or less, and there is a two-mile speed limit stretch through Vermilion Bay. Our company truck is a high-miler GMC panel with a six-banger. The valves must have been floating as Ahmed blew past the Viking. No wonder he no see!

It must be stress, or maybe my brain is suffering a synaptic seizure, but I draw Ahmed a map showing the exact location of the Viking, add two quarts of oil to the tough old GMC and turn Ahmed loose again, with the admonition to slow down, for heaven's sakes!

I lay down on my bed to rest my aching head and Merle Haggard is singing, "I want to settle down, but Ahmed won't let me." And he won't.

Twenty minutes later Bill knocks on my door. Ahmed is on the phone, and wants to talk to me. What now, I wonder? I go to the phone and Ahmed tells me the truck is broke. Holy Cow! The little jerk has blown the motor for sure. "What do you mean, broke," I ask?

"I don't know – maybe the tire is flat."

"Well, change it, Ahmed, dear Ahmed."

"I don't know, maybe jack don't work."

Well, Hell's bells! Ahmed tells me the truck is just a little way past Vermilion Bay and he is calling from the Esso at the junction of the Red Lake highway. I tell him to go back to the truck and wait. Mel has come out of the bush, I fill his pickup with gas and we go to the rescue. We arrive at the scene to find Ahmed making no attempt whatever to change the tire – the jack is still in its place under the hood. I am finally realizing this little bugger's mind-set. He is of the ruling class, and we are less than camel dung to him. He doesn't change tires – unclean people change tires, so Mel and I change the tire and I take the truck and Ahmed back to camp.

On our rescue mission, I noticed the shoulders on both sides of the highway showed signs of disturbance, and I thought I had seen tire skid marks across the pavement. I query Ahmed a bit, but he just shrugs. "Maybe bus go by."

The next day I take the GMC to the Esso station and put two new tires on the back. The Esso is also the bus stop, and as the guy installs my tires he says, "This looks like the truck the Greyhound driver was telling me about yesterday."

I ask him what he means, and he tells me that the driver told him that a vehicle of the same description had slid sideways across the road in front of the bus, missing him by five feet. That's it – Ahmed is grounded.

I call Don before I leave and tell him I've had it with Ahmed.

"Then fire him," says Don.

"If I hire them, I'll fire them," I tell him. "You hired him, so you can fire him."

"I didn't hire him," says Don.

"Then who did?" I ask.

"Ralph hired him," says Don. (Ralph is Noranda's chief geologist, in Toronto.)

"Then Ralph will have to fire him." I hang up and go back to camp.

On the way back I do some mental re-evaluation of my predicament: Ken has been chugging away on the mag, and I have been to Port Arthur for stuff such as snowshoes, a vertical loop unit, and a double-track Skidoo. We have some vertical loop work to do on the property, and being November, the snow cover is building up. We are not quite on the snowshoes yet but one more good snowfall will do it. I decide to give Ahmed one more chance before snowshoe training. Pointing the vertical loop doesn't take much brain power, and perhaps he can handle the concept. The next day I load the unit on our Skidoo sled and haul it and Ahmed in to a set-up point on the claim group.

There is an obvious fault running east/west through the centre of the property, and I will test it for possible conductivity. Beavers and Mother Nature have created a small lake at a widening of the fault, and I will start there. I will be reading on the ice for a few hundred feet, and perhaps Little Ahmed will get the hang of it. I am not doing a search square, and will only be reading eleven stations per line.

I set up the unipod without much help from the squirt. With little snow on the ice, I have to cut holes for small stakes to tie the ropes to. Once in a while I give him a rope to hang on to – Ahmed doesn't move much anyhow. I show him how to start the engine, and tell him to point the coil at me when I take the readings. I tell him that when I hit the bush he is to orient by sound, run the motor for 30 seconds, shut it off and wait for me to holler again. Nick's raven has skewed the lines so I can't use the orientation board. Ahmed "no understand" that anyway, I'm sure.

The lake part goes OK, but when I reach the first bush station, the motor is still running and I can also tell he is not pointing the coil at me. There is no point in hollering at him, he can't hear me over the motor noise, but as I walk back to the set-up I scream every obscenity I can think of until I exhaust my repertoire, and by the time I am back to the transmitter I am calmed down a bit. Ahmed seems a little nervous. He keeps his distance.

OK – I'm from a family of teachers, so I give Ahmed a tutorial. I cut a good-sized branch and use the lake as a blackboard. I draw an overhead view of the fault, with our set-up and cross lines. I draw a picture of a fault, and tell Ahmed to imagine he is 200 feet underground, and facing the fault. Then I draw primary fields and secondary fields – the whole schmozzle, talking all the time. By the time I finish, I have used up 100 feet of white blackboard and it's full of circles, swirls and hieroglyphics. I'm sure Stephen Hawking or Einstein couldn't have done it any better. I turn to Ahmed and say, "Do you understand? And his reply is, without a word of lie, "Jesus Christ, I wish to **** my mother!!"

I am literally floored! I fall on my back on the ice, look up at a clear blue sky and imagine I am in a happy place, as I make a snow angel.

I get up and tell Ahmed to "Sit on the double track, and don't you dare move a muscle, you little shithead!"

I have to work off my anger. I know pretty well where my stations are, and I use my instruction wand to poke into the snow to hold the transmitter coil in place. I start the motor/generator and leave it running and trot out to take my reading, then back, move the coil and trot out again, and so on. The JLO two-cycle is a fuel miser, and by the time it runs out of gas 2½ hours later, most of my anger has dissipated. I have trotted about five miles to take 2200 feet of readings, and I'm whipped. I tear down, load the sleigh, chase Ahmed off the double track seat and peel out for camp without a backward glance. I hope Ahmed the

Idiot has to walk home, but no luck. When I get back to camp he is hanging on for dear life. I don't even have my supper. I just crash. But my mind is made up – Ahmed must go.

The little freakaziod actually makes it easy for me. For quite a while now, I have noticed that there seems to be a build-up of ice on our little plywood step at the cabin door. I assume that Ahmed is washing up and is too lazy to throw the water to the side, and I tell him to stop. Of course, he shrugs.

The next day is D-day, and I go for my 6 am breakfast. As I walk back to the cabin I see a stream of liquid arcing out and splashing on the step. Dirty Dying Dinah! That scrawny spawn of Satan is pissing through the screen door! When I reach the cabin, the door is closed, but I kick it open and storm in. Ahmed is standing on his bed in the corner, heading for high ground.

All I can say is, "Pack your bags, they need you in Port Arthur."

I make sure Ahmed gets on the 10 am bus and as the Greyhound pulls away I have a feeling of euphoria. Ahmed the Terrible is gone, but never forgotten. (Two years later I am at Pickle Lake during the Umex rush, and I overhear a conversation at the bar. They are laughing about a Umex geologist who never goes into the bush, and runs around all day in pyjamas – and my blood runs cold!)

I have hired another Ken- Ken Fisher – a down-home Emo lad. Kenny and I break him in on the JEM and he catches on immediately. Most men do – the JEM is not rocket surgery. I put the two Kens on the JEM and I take over Kenny's mag - I want to be alone. I have a bit of an Ahmed hangover, and one day, for some forgotten reason, I holler at Kenny. He just looks at me and tells me to quit acting like a baby. He's right – I am being childish. I go out to the back yard and eat worms.

Sidebar: In retrospect, I have to admit that Ahmed helped out quite a few people. Over the next few years, I would hire quite a few men, some better than others, and I seldom fired anyone. This type of work is not for everyone, and some will try it for a season or two and leave to find a different niche in life. You roll the dice and count the dots. What can I say?

On December 12 we shut down the job for Christmas. We have finally started make some steady progress on the grid and will return in January. Kenny and Ken head homewards. I spend the day storing gear in the cabins. I will drive home tonight, spend a day with my family and then go on to Port Arthur with the maps and the double-track Skidoo. We are trading it in on a new one.

I am almost ready to leave, and I have to load the Skidoo by myself – there is no one else around to lend a hand. Now, a single track is awkward to load without a ramp, but a double track is actually quite easy.

The front end is very heavy, but once you have it on its tail you sort of walk it side to side to the truck tailgate and drop the single ski onto the truck. Then you simply pick it up by the ass end and slide it in.

With the Skidoo loaded, I am ready to go, and I head for the restaurant to have one last coffee and to wish Bill and May a Merry Christmas. As I sit at the counter I think about the last couple of months and suddenly I sit straight up and give my head a shake. Eleven months ago I had been off work nursing a bad back, and my sawbones had told me that if the problem persisted I should consider a spinal fusion. Now I had just loaded a double track Skidoo without a twinge. A month of strike followed by daily jaunts in the bush has cured me! The problem wasn't my back, it was the paper mill. I will never have back problems again!

Chapter III   Winter '69

Wrapping up Bridges Township

Good old January 1 rolls around and I leave home again. Why is it always Jan 1? What's wrong with 3 or 4? One of the mysteries of the bush, I reckon.

I pick up Kenny and the new double track Skidoo in Port Arthur and head for Bridges Township where Ken Fisher will meet us tomorrow.

With no distractions now, the job goes smoothly – too smoothly, in fact. With two good men on the job, production settles down to a steady pace. There is lots of snow and we are on snowshoes now, and although there are a couple of high ridges on the grid, the lines are well laid out and the bush is mostly poplar and birch without a whole lot of thick underbrush. There is a fair amount of outcrop on the property and some of the smaller ridges are a bit steep. You can always find a way around a sheer wall if the grade is not too steep, you simply grab hold of small willows or a tree branch and carefully climb the six or seven foot slope, your snowshoe webbing acting as snow tires, more or less.

I am still on the mag and am a day ahead of the two Kens and their J.E.M. So I am breaking trail for them. Now, I am not a keen observer of the human condition, but I know that Kenny, (who is always lead man) will see that I have made it up the hill, and if I can do it, so can he.

So I set traps for him. As I climb the little hills I take time to break off the little bushes I have used for handholds and replant them in the snow, enticingly close to my snowshoe trail. Kenny will come along, reach out for a handhold to help pull himself up the hill, and of course, the branch pulls out of the snow, and Kenny topples backwards to crash and burn.

At suppertime I put on my best innocent face and ask Kenny how things are going in his world, and Kenny is mystified. He says he can't understand why I can make it up the hills and he ends up on his bum about three times a day.

It has become our habit to go to Vermilion Bay every Friday evening. We can do a bit of shopping and then have a couple of cool ones and a game of shuffleboard. After a week or two of setting Kenny-traps I fess up on Friday night, and Kenny calls me a bunch of bad names, laughing so hard he falls off his chair. You see – if things are going too smoothly, you always suspect something really bad is going to happen, so setting Kenny-traps is like being inoculated for smallpox. A little irritation now can prevent bad things in the future.

Another problem can arise when the job goes well. We can finish too soon. This is a good set-up here, with good food, a short distance to the grid and a town close by but not <u>too</u> close, if you get my drift. If we wrap this up, where will we go next? I ask Port Arthur, and Don avoids the question. All I know is that we will be spending next summer at Red Lake, so I don't push the boys, and Don doesn't push me.

I mean it when I say this is a step up from an isolated frame tent camp. When you compare the logistical cost of this grid with a remote fly-in job, the cost benefits to Noranda are considerable. Add that to the availability of a hot shower, a laundromat 17 miles away, and evening TV in the restaurant, and everything comes up sevens.

I spend a lot of evenings drinking coffee and chatting with the owner. Bill is an old trapper and commercial fisherman who has done his share of hard work in the past. He had his eye on this location for a few years, and five years ago he decided he wasn't getting any younger or any smarter, so he put his savings and a bank loan together and built the place himself. It was a tough row to hoe at first, with the winters being pretty skinny, and Bill tells me this story.

Table Scrap

In their third year, Bill and May were just starting to see some black ink. A logging contractor from Pinewood in the Rainy River Valley had 2000 cords of pulp to move from the nearby Gordon Lake road to the mill in Dryden. He arranged for room and meals with Bill, and also would fuel his trucks at Bill's pumps. This was still pre-diesel days, and those big 409/427 engines were gas-guzzlers. Bill welcomed the business – it was a great wintertime bonus.

Of course the deal was this: Bill would invoice the logger at month's end and receive a check, so Bill trotted into Dryden to talk to his bank manager and fuel distributor – and they were unimpressed. The bank refused to give Bill a line of credit, and the Gulf guy insisted on cash or no gas drop. By mid-January, Bill was up against it. The logger dropped by for coffee and Bill had to tell him that in three days his tanks would be empty. The logger would have to buy his gas elsewhere.

The guy went to his truck and came back with a cheque. He had figured out how much gas he would use in two months and he paid Bill in advance. Bill told me he almost kissed the guy. The very next day Bill had a load of gas and a new bank manager.

Shortly after hearing Bill's story I asked Don at Thunder Bay about Noranda's bill paying policy and was told that invoices were always paid on receipt. I liked that, and in the next two years I would seldom have trouble arranging rooms, groceries or flying, on credit. I had to use my own credit card occasionally, but since I was reimbursed much more slowly than Joe McBusinessman, I did so reluctantly.

Sidebar: You know, banks are rarely a friend to small business. Canada's strength has always been in the hands of honest, common-sense people like that Pinewood logging contractor, and in my experience I have to include most mining companies in the logger category. They can be ruthless when dealing with each other, but they generally look after the little guy.

Back to work, more or less. We are almost finished the mag and J.E.M. and are using the double-track to access the eastern part of the grid. Soon we will be on the vertical loop and we are nearing the end of our stay here. One day we awaken to find that more than a foot of new snow has fallen, and as we head out that morning we see Bill with wrenches in hand heading for his snow plow. We stop to find out what's up. His plow is busted. Bill uses an old Allis Chalmers grader to clear his parking lot and driveway, and it has quit on him. It's not much more than a modified farm tractor with a grader blade under its belly, but it usually does the job. Today it won't run – the magneto is fried and Allis has no spark. Bill has a new magneto coming in on the ten o'clock bus and is going to remove the old one. As usual, a cold front has moved in following the storm and it's -30 outside – daunting weather for pulling magnetos (or pulling instrument readings, for that matter.) I tell the two Kens to put the sled away and take the day off. I'm going to help Bill.

So we unscrew nuts, blow on frozen fingers, warm up hands with numerous cups of coffee and by 2 pm we have Allis chugging. Bill clears his snow and says "Meet me at the bar." He is going to buy the beer tonight. I take the two Kens with me, pointing out that those two wretches are undeserving beneficiaries of my hard work. Are they humble? Yeah, right – they think it's pretty funny. It's Friday night, the bar is hopping, and Bill gives us the royal treatment. He also treats himself pretty good, and we follow him home to make sure he gets there.

The next morning is cold in more ways than one. Bill had hit the Co-op before he hit the bar and he left the groceries in his truck for five or six hours. The veggies are frozen solid, and May is a little on the cool side herself! Bill has to go back for more groceries, and for the next week the two Kens and I have to eat snowbank potatoes while all around us the other customers are enjoying fresh french-fries.

Table Scrap (Courtesy Brian Tierney)
Bill could have been an Honorary Bushwhacker. For us guys it was routine – groceries before refreshments.

It might have been in the '70s. It was a summer camp, fairly large, with road access and a cook. Henry "Pass Me Da Hax" Levac was in charge.

Henry went to town for grub. Along with the groceries, he threw a few blocks of ice in the truck for the cook. It was a hot day, so he stopped in for a cool one or six. Back at camp he unloaded the truck.

"Hey," said Henry, "Somebody stole my hice!"

We are done the mag and I.E.M. and so we load up the vertical loop and all three of us head into the little lake to start the final leg of our survey, and immediately get stuck in the slush. Either the heavy snowfall has pushed the ice down, or the beavers have raised the water level, but for whatever reason, there is now six inches of unfrozen water on top of the ice, but under a foot of snow.

If you are cruising along on an unfamiliar lake, you learn to watch for signs of slush. You avoid darker looking areas of snow, stay away from creek mouths, and keep your speed up.

You also pay attention to your backtrack to watch for darker sections to be avoided on your return trip. So much for slush avoidance in theory: In reality, with a double-track with two up and a third man on a heavily loaded sleigh, when you hit slush you are stuck – no two ways about it. Now the fun starts.

First you jump off, unhook the sled, and with all three of you pushing, try to get the Skidoo up onto firm snow. But, the tracks are spinning in water, frozen slush builds up on the single ski, and (in my case) your moccasins are already soaking wet. Plan B swings into place.

We strap on our snowshoes, and while two of us go to cut small poles, the third man packs a trail veering towards shore to find better going. We then lift the Skidoo, slide the poles underneath, and with two men holding up the back of the machine, spin the tracks to get all the crap out of the tunnel before it freezes and immobilizes the unit.

Now we cut many spruce and jackpine boughs. Some are used to cover the snowshoe path to safety, others serve as a pallet to hold the vertical loop gear which is off-loaded. While two men turn the sleigh over to clear the ice build-up from the runners, the third man drives the Skidoo ahead onto the spruce bough trail. A rope is attached to the sleigh and everything but the EM gear is moved to higher ground. We are fortunate that camp is nearby. We can go back for dry footwear, and we do so – it's lunch time anyway. This afternoon we will return to rescue our vertical loop. The crash-and-burn scene will be frozen by then, and we will stash the gear near shore until tomorrow. The rest of the day will be spent snowshoeing the lake picket lines to pack trail in order to prevent further slush incidents. Fool us once, Mother Nature – your fault: fool us twice – our fault.

Sidebar: I'll take a little time here to rag on the snowmobile industry. By 1969 things were already getting out of hand, with more horsepower, ever more horsepower. The most actual work I have ever seen produced by these handy inventions of Armand Bombardier was that little tin seven horse in Chibougamau. Our twelve horse long track jobbies were no slouch, either. With one up and the back end sinking in slush, it would keep on climbing onto fresh snow as long as you held her wide open on the throttle – likewise with our ten and twelve horse double tracks. But now we have to have high-output/twin cylinder models and the weight is on the front – not good for slush and difficult to rescue. Bombardier got back to basics a few years later with the Elan model, but it lacked the longer track of the earlier

Skidoo. Maybe I'm just an old curmudgeon, but dang it anyway, why couldn't they leave well enough alone?

Table Scrap

Further to getting slushed in – at least we had only six inches of slush to deal with.

In January of '71 I was staying in the Savant Lake Hotel and was sending a three-man camp into the north end of Savant Lake itself. They were hauling the camp gear in by Skidoo and were accompanied by the line cutting contractors who had their own Skidoo. They would camp side by side and my men would be following on the heels of the line cutting crew.

I was in my room going over some maps when I heard some hallway activity, but I thought nothing of it. About ten minutes later the landlady knocked on my door. "You better check on your goofy crew," she says. "They're all sitting in the bathtub with their clothes on."

I head for the communal bathroom, and five guys are sitting on the edge of the big old claw-foot tub, laughing like crazy. They had hit two feet of slush, and by the time they dug out they were solid ice to the knees. They had to soak their feet in hot water just to get their boots off!

Table Scrap: On Beavers, Beaver Dams and Beaver Dam Decommissioning.

We clumsy yard apes know that beavers are dam fine engineers. Some of us have seen the structures in situ, others have learned from the many fine TV docs, but until you try to dismantle a beaver dam you have no concept of how tightly they are woven.

That little lake in Bridges Township was a typical deal. Not much of a pothole, but being a wide spot in a shear zone, it invited beaver dam modification. In 1969 it was on its last legs as far as natural beaver habitat – the surrounding broadleafs would need to regenerate for a few years but it was evident along the shoreline that the lake level had been quite high in years past.

The shear zone valley was separated from the Trans-Canada Hwy by a ridge, and the little creek at the east end tumbled down a rock face 50 feet or more to join a larger creek which then crossed under the highway. The main dam had blocked the creek at the top of the rock face and we could tell that it had once been quite a structure. While doing our vertical loop EM we used an old tote road, passing by the old dam every day.

Three or four years after we worked the Bridges claim group I was living in Vermilion Bay and met Ray Hertz. He was the bush boss for Tri-Lake Timber, a Kenora based logging outfit. I happened to mention the tote road and beaver dam – Ray had a story for me.

The old tote road was used to forward logs out of Bridges to a landing near the highway – saw logs for Tri-Lake's mill and pulp for the Kenora paper mill. They had to cross the creek near the dam and the doggone beavers were messing with production. Tri-Lake could never count on the ice being consistent and by spring melt time they still had logs to move and four feet of water to deal with. Something had to be done.

Try tearing a gap in a beaver dam sometime. It's hard, frustrating work. You can breach the dam but by sunrise the next day the gap will be filled.

So you try dynamite. This is also hard work – it's not easy to penetrate the dam to get the charge deep enough to blow a large hole. A large old structure has so many interwoven sticks that the blast is dissipated and little damage is done, so you go to Plan B.

Background: Tri-Lake Timber was owned and operated by the Hertz/Herbacz family. The H/H clan were Lake of the Woods settlers and back in the day Kenora (Rat Portage) was a lake transportation hub. Before rail took over there was at one time 40 steam freighters plying the waters to Fort Frances and Northern Minnesota. There was lots of stuff from the old days still kicking around and back in the sawmill yard was a <u>huge</u> old boat anchor – a treble hook dealie. Maybe this would solve their beaver dam problem.

So they haul in the anchor and with some difficulty insert it in the dam using a small cat to tramp two hooks as deep as possible. They hook it up to the cat and it would not budge. A D-7 sits nearby and is brought in to play.

The plan was to breach the dam, lower the water level to a workable depth, tote wood around the clock and then leave the beavers to do their patching. The D-7 did the job – too well.

The dam dam came out in one piece peeling back like a huge band aid! The whole lake went down the hill and washed out the Trans-Canada Highway! The main transportation link joining Eastern and Western Canada was out of commission for 24 hours.

Tri-Lake took the 5th and the beavers took the fall.

Table Scrap: On Beaver Dams and Bush Art (with added refrigeration)

Ray Hertz also told me that I should talk to his great uncle, who had done a bit of prospecting in the old days, so one fine fall afternoon I stopped in at the H/H family compound in Kenora, and met the old gent.

He would have been in his mid-eighties and was born around 1890 in a log cabin in Rat Portage, the first of the clan to arrive after they settled. We never did bang a rock off an old outcrop that day. We just sat in the afternoon sun near his comfy bachelor domicile and swapped yarns. As usual, I was in a trade deficit position. At 35 I had little to offer and well aware of my lack of whiskers, I just listened and learned.

He was an artist – self-taught and uniquely so. "I paint with wood chips," he said, and brought out a recently completed nature scene.

It was two feet wide, eighteen inches high and neatly framed with rustic barn-board. It depicted a small bay surrounded by mixed tree species and the sky was sort of an overcast whitish blue. Not one lick of artificial paint anywhere – not one lick – it was all done with wood chips glued to a stiff backboard! And here's how he did it.

He haunted beaver dams – old and new. He collected chips left at the base of trees the beavers had knocked down – some fresher than others. He also pulled out sticks from older dams and made his own chips from these – this was for added colour.

Let me explain, keeping in mind that I am not a geo-scientist,

You can cut a sapling – any sapling. It may be poplar, birch, tag alder or conifer, but it is just a tree, and inside it is tree-coloured.

But use it to build a beaver dam, and it changes character. It will be mostly underwater for years, and will decompose ever-so-slowly. The sap will leach out leaving natural shades of gray and blue – teal, azure, deep blue – all colours of the blue-green spectrum – and here's why.

All glacial till holds traces of our world's chemicals, rock types and such and the trees bring these traces into play as they use ground water in their growth process. Soil sampling bypasses the tree storage system, but back in the day there was some tree sampling done.

This was generally known as "twig-o-chem", and was not your average prospecting tool due to the high cost of lab work. However, it would give an idea of economic metal percolation and, like following float rock to a source, might establish a glacial trend. Got that?

So the buried sticks in the beaver dam retained that chemical and the bornite or cobalt remained to give Mr. Herbacz his palette of colour.

Chips became lakes, became tree trunks, dark green conifer boughs, lighter green deciduous leaves, shades of white/blue slightly overcast sky and all natural, all with beaver tooth marks and all so beautiful.

His only concession to our modern world was the glue and the clear urethane to protect his masterpiece. Wonderful!

And then, on the typically warm, sunny, fall afternoon, the subject veered to refrigeration – how so, I do not recall.

They lived on a little lake and I assumed they put up ice in the winter. "No," he said, "We built a walk-in refrigerator.

First they dug a pit, 8'x10'x6' deep, filled it with water, and left it open to freeze solid in the cold winter months. They built a log blockhouse big enough to cover the pit including a door insulated with sawdust. Before spring they pulled it over the pit with a team of horses and covered the ground around the base with two feet of sawdust. The plank floor would allow the cold to percolate upward and the ice would last until October. Then they would bring back the team to remove the house and restart the cycle. Zero environmental impact!

We find only two conductors on the Bridges property. One follows along the fault zone and the lack of frequency correlation and mag association on the east end of the conductor indicates a water filled shear zone. The west end is more interesting and will justify a follow-up drill hole next summer. (The drill intersected an inch of lead/graphite, and the claims were dropped.)

On the last line of the last vertical loop setup, we find an interesting little blip on the edge of the northwest corner of the grid. I trace it onto open ground, stake six claims and put in a compass grid, completing the survey with J.E.M. and mag in addition to the vertical loop. I write a report recommending drill follow-up, but my report is treated with casual indifference by Port Arthur and Toronto. I file this spot in my memory bank.

Bridges Township is wrapped up. Ken goes home for a few days, Kenny and I load the Skidoo, sled, and various pieces of survey gear, say a fond farewell to Bill and May, and head for Port Arthur with the overloaded GMC hugging the pavement. I spend two or three days tuning up maps in the office and go home myself for a short break. I will return to Port Arthur to do one last job before break-up.

Sidebar: It took me a couple of years to figure it out, before I came to the conclusion that the Bridges claim group was a combination make-work and a reward to an old prospector. To Noranda's credit (and a bit of a throw-back to earlier days) they kept two prospecting crews in the bush. Like the two old boys on Frotet Lake in '61, they spent their summers prospecting and their winters on retainer. Two of the prospectors (Chic and Leo) had been with Noranda since 1947 and the Bridges property was prospected by George Cates, who had been with Noranda for more than ten years. Part of the deal with these guys was that if a property was brought to diamond drill status then a bonus clause kicked in.

By the time we had finished the geophysics on the Bridges property, Toronto had already lost interest. The only reason the drill was brought in was because the target's proximity to highway and camp made the logistics very cheap, and gave George a chance for a bonus. This also explains their reluctance to move the drill three miles farther and check out my conductor. Their heart was in the right place. It was their brains that were screwy.

On April 17th I returned to Port Arthur with Ken and a new hire. Gene cuts wood for a living and is shut down over break-up so he will be temporary. We just need a fresh body for a short time. Kenny has made it quite clear that he is not going back to work until the summer season.

We should have stayed home also. Even though the winters were longer back in the '60s, by April things could start to get messy, but the company wants us to survey a 20 claim block before break-up, so off we go to Renabie.

Renabie is a little lumber mill town 30 miles east of Wawa, Ontario. There had been a small gold mine nearby at one time and since the '80s mining activity has been on the upswing in the area. Nowadays highway 101 connects Wawa to Chapleau and Timmins, but in 1969 it was strictly a timber access road, with a seven mile branch running into Renabie.

The grid is right alongside this access road, so on our way in we stop to check it out. The area had been clear-cut a few years before and has grown back in small balsam fir and poplar. There is still plenty of snow cover remaining, but today it is melting fast. We head into town to check out the hotel, but it is not easy to get to. As we turn a corner one mile out of town, we are faced with a mile of road that can only be described as a quagmire – nothing but mud ruts filled with water, far beyond the capability of our two-wheel-drive GMC. We stop to eyeball the situation. A bulldozer is idling by the side of the road and I get out to talk to the operator. He has no English, but I point to town and he pulls the cat around to the front of our truck and points to the chain hanging on the drawbar. I get down on my knees in the mud and hook the chain on the GMC frame and he hauls us the last mile. In town the streets are a little drier, so I tell the guy "Merci," unhook, and we go to look for the hotel. Things don't look too good right now, but we soon find out they are not going to get any better.

The hotel guy speaks my language somewhat and has two rooms ready for us. The problem is, he refuses to invoice Noranda – he is a Johnny Cash kind of guy and cash in advance is the only acceptable route I have. Fortunately, I have enough in my pocket to cover two days' lodging and meals. I will have to whip into Wawa on Monday morning to straighten things out with the Port Arthur office. To say I am unimpressed with Don at the moment is putting it mildly, and I am looking like a doofus in the eyes of my two men.

What the heck, it's Saturday night, and after supper we retire to the bar for beer and shuffleboard.

This is obviously a très Française village. Everybody speaks French, and the TV is on the French channel. No problem – we sip our beer, shoot a little shuffleboard, and discuss our plans for Sunday. I tell the guys we will hit the bush at daybreak, counting on the overnight frost to tighten up the mud and make it passable in the early morning. We will work until 2 pm and then hope the bulldozer will get us back to town. I talk with the bartender and find that my plan is not going to fly. The bulldozer doesn't work on Sunday. In fact, the cat operator is pretty much on the job by appointment only. It seems that the mill is shut down until mid-May, and the only reason we did make it into town was because the dozer guy was waiting for the last Saturday shoppers to return. Holy cow! We're stuck in Lodi again!

We are pretty sure we know where we are at by now, but just to show that the coffin always needs another nail, the hammer drives one in an hour later. The bar has filled up, and it's Hockey Night in Canada. Of course the Montreal Canadians are on the tube. I'm a fan of Guy LaFleur, Jean Beliveau et al myself, but it's all in French. We decide to have a game of shuffleboard and drop in a quarter. Hmmm, the scoreboard doesn't light up. I study the situation and see the plug laying on the floor. I assume it has fallen out and I fail to see the glowering disapproval on the faces of the bar patrons as I crawl under to plug it in. The scoreboard lights up, and as I go to make my first shot, the bartender hustles over and pulls the plug. "No shuffleboard when les Canadiens play!" he says, and stomps off. We are pretty unpopular in Renabie. I sense that trouble may be brewing, so we go to bed early.

No problem – early Sunday morning we head out over the frozen mud, but not to the grid. We have our packsacks in the truck, and happiness is Renabie in our rear view mirror.

Chapter IV   Summer 69

Red Lake
I spend an enjoyable month at home. This is a big plus working for Noranda. They are not as regimented as Inco regarding earned time off. We always take a month off in the spring and at Christmas, and in fact Don had urged me to spend more days at home the previous winter. There were a few reasons why I failed to do so. For one thing, I had no one I trusted to leave in charge, and even if I had left the two Kens alone for a few days at a time, they would not have had any transportation.   Kenny is also a family man. If I go home every two weeks, why should I expect him to stay on the job? This problem will trouble me for a few years – job versus family. Eventually it will be a factor in my divorce, and I am too stupid and too wrapped up in my bush career to recognize the signs.

In mid-May I give the family a smooch and head for Port Arthur to spend a few days preparing for the upcoming season. It's shaping up to be a busy one. In 1968 Selection Trust of England (SELCO) had hit a copper deposit in the Uchi Lake /Confederation Lake area, 40 or 50 miles east of the Red Lake gold camp and a staking rush had followed.
Serendipitously, Noranda had McPhar Geophysics fly the area just before Selco made their announcement, and our contract stakers and line cutters had been busy all winter. This summer I will be the go-to guy on the Red Lake project and I welcome the challenge

I will have two geophysical crews. One will be surveying the various anomalies – usually blocks of twelve to twenty claims. They will be moved from job to job by float plane and will live in easy-to-move fly tents.
Noranda has also optioned 150 claims from a group of Red Lake prospectors. This block is 90 air miles north of Red Lake on Black Beaver Lake. It is your typical non-descript Shield lake, five miles east/west and up to four thousand feet wide with a boring, solid tree shoreline.
Here we will build two plywood floored frame tents. This camp will sleep up to five men – four in the bunk tent and an overflow cot in the kitchen. The four men staffing Black Beaver and the three-man anomaly-chasing crew, will be under my direct supervision.
Another three-man crew will be moving from grid to grid taking soil samples. Although not working directly under me, I am responsible for their well-being, so I will be keeping a close eye on them. They are all students with little bush experience, but are good young fellows – always willing and able to handle most of their foul-ups. They are determined to be successful while getting the most out of their first summer on the Shield. Every day dawns bright and clear for these young pups.
There is a three-man line-cutting crew completing the eastern part of the Black Beaver grid. They are camped on a little pothole just east of Black Beaver and will be finished in mid-July. I will only visit them once to deliver some work assignments. They are self-sufficient and will be no problem.
Two other camps will get some attention from me. They are two-man prospecting camps and although also self-sufficient, I will have to contact them from time-to-time.

Add in two roving geologists. They are of no fixed address – hopping from one grid to another. One will not stay long in Red Lake – Noranda always has showings to check out elsewhere. The other is a permanent hire. He is a recent graduate from Haileybury School of Mining Technology and has spent three previous summer seasons with Noranda. He is a pretty good guy and helps the junior boys a lot.

Sidebar: Finances had been pretty much "no frills" for that young fellow during his Haileybury years. He told me that with his first check he was going to buy himself a wristwatch!

Now, in 1969 you had two choices of affordable watches – wind-up or self-winding. Windups had to be serviced daily, otherwise you had to use the sun or hope your radio worked. Self-winders had their own drawback – pounding rocks or swinging an axe would over wind them and time would stand still. Put them in your pocket and they might wind down unnoticed – problems, problems.

So he was going to buy a Bulova Accutron – 200 bucks! – half of his first freedom check!

Today one can pick up a battery-powered bang-on-the-second timepiece for $29.95 –how time has flown!

So that gives me 16 men and six camps to keep tabs on. Someone has to know where they are now and where they are going to be next. In normal times, any competent air service will keep track of camps they have dropped off. But in a busy year such as this one is shaping up to be, mistakes are inevitable. To underline how easy it is to lose men, consider the following:

Table scrap

Pre-1969, two prospectors from Port Arthur flew into their claim group north of Nakina, a railway town on the C.N. line situated near the top end of Lake Nipigon. While I know the name of the air service they used, I will withhold that info for obvious reasons.

The two prospectors, Gus and Ozzie, planned to spend a few days trenching and sampling their showing. Spring break-up was right around the corner and they wanted to bring back some good samples to drum up interest with a mining company, hoping to option the property.

These boys had spent their life in the bush, and although they packed light, they took enough grub for ten days, well aware of changeable spring flying weather. The pilot dropped them off, returned to base, threw his suitcase into his car and left for Florida for three weeks of fun in the sun.

Ozzie lived with his elderly mother who was used to his uncertain schedule, but after three weeks of no Ozzie and no news, she went down to the Waverly Hotel bar, the local hangout and news central for all Northwestern Ontario exploration activity. "Has anyone seen Ozzie?"

A few caps were lifted and a few heads were scratched. "Why no," opined one old gent, "Haven't they got back from Nakina?"

A phone call was made forthwith, and at the base the freshly-tanned pilot overheard the conversation between Ozzie's mom and the base manager. "Holy Cow!" said the pilot, "Nobody picked them up?"

The planes were off the lake being switched to floats. A chopper was sent in and the two prospectors were pulled out – 10 days with no food and 40 pounds lighter – and that's why I kept tabs on everyone.

Here's the real kicker. Gus and Ozzie paid for the chopper! A mining company had a financial interest in the air service and the boys didn't want to rattle their chain. Go figure!

Our first order of business is to build the camp at Black Beaver Lake. Before leaving Port Arthur we load a couple of new canvas 14x16 tents into the GMC, add a whole bunch of camp related equipment, tie a canoe on the racks and with Kenny and I in the front bucket seats and two new hires on top of the load in the back, we head for Red Lake.

It's a long seven hour haul to Red lake and I feel a little sorry for the two guys perched on their sleeping bags, but not sorry enough to give up my seat, nor is Kenny. The two new guys are Rob and Tim, both summer students, and this is Rob's second summer.

We arrive in Red Lake, check into the Red Lake Inn and have a late supper. The Inn is a big old two-story hotel on the main drag on the east side of Howey Bay. A set of wide steps leads up five or six feet to the front entrance and on the first floor is a comfortable lobby and reception desk. There is a good dining room and a decent sized ballroom (should anyone decide to have a ball) and on the second floor are old-fashioned high ceilinged rooms, quiet and comfortable. Beside the front steps, another set leads down to the bar, essentially in the basement of the hotel, which is why the second floor rooms are quiet, which the bar is definitely not! To locals, the bar is known as the "Bucket of Blood" or simply "The Bucket." Not that they allow things to get out of hand, but on Saturday nights things tend to spill out onto Main Street and in 1969 every night is Saturday night.

Across the street from the Inn is the Annex. It is sort of a motel type building, part of the Red Lake Inn and built on the shore of Howey Bay. My room faces the bay and this will become my usual hang out for most of the summer.

Believe me – this town is buzzing. It's not enough to have various exploration crews and stakers coming and going. Oh no! Red Lake has decided to go modern with their sewer and water infrastructure, and the town is full of construction crews digging up the streets and blasting rock for water and sewer line installations. Rock is one thing Red Lake is not short on.

To add to the mix is the fact that Red Lake is the main hub for a number of Indian reserves (nowadays called First Nation communities) and remember – SELCO is in the early days of developing their Uchi Lake ore body. An all-weather road is under construction from Ear Falls to Uchi, but for now Red Lake is their main supply depot.

The next morning we head to a charter air service base a little further up on the west side of Howey Bay. On the way, we pass Green Airways and a Shell fuel storage site. Just past this is the Ontario Central Airlines (OCA) seaplane base. We will be using OCA for most of our flying this summer.

I meet Dave Harvey, the OCA base manager, and my first impression is that this guy is right on top of things. He is running at least 10 float planes out of Red Lake and besides the mining activity, he has tourist camps and daily sked flights to at least three of the numerous reserves around the north country. He has three twin Beechcraft on floats, three Beavers, two

Norsemans, and probably a 180 or two. OCA also has a base in Kenora, and aircraft are always coming and going. I certainly would never be able to keep track of them, but Dave does the job with ease. He has four hands and ears it seems. One hand holds the 2-way radio mike, another points out the next plane to be loaded, with instructions to the dock boys (his office fronts the main dock), another answers a phone which is always ringing, while his fourth hand shakes mine and he'll be with me in a minute, as he swivels to his wall map to stick another pin in to mark a new camp location with his fifth (?) hand. It's pretty to watch, and I am mesmerized.

Twin Beech on floats

As I stand there like a dummy, two twin Beeches are already being loaded with our camp lumber and other material. Port Arthur has called in the order to the local lumberyard and Dave has made the necessary arrangements for delivery to the dock. The initial grocery order is on its way, and we will be going in with our personal stuff on a Beaver. Dave passes me on to the head dock boy and he shows me our storage area in an attached warehouse. Here, OCA has bins sort of like horse stalls, with each one designated to a customer for temporary storage. Some stuff will wait for the next flight and some has come out from another camp to overnight for tomorrow's flight. A few propane cylinders are lined up along one wall with "Dave's LLLLL" written on them with magic marker." What in the dickens is Dave's LLLLL?" I ask.

"Easy," I am told "It's Dave's Little Long Legged Lake Lodge. Everybody knows that."

Well I guess I'm now to be considered part of everybody.

It's hard to describe the complete Red Lake scenario with a few words printed on white paper. It should be chaos, but it is not. You would expect frenzied activity and feverish excitement. Instead, you have many people doing many jobs at an efficient steady pace. If your job is managing a seaplane base from the crack of a 4:30 dawn to 10:30 p.m. moonrise, you do it. Winter is only months away – you can catch up on your sleep then.

Side bar: Red Lake was born in the mid-twenties, when word got out that gold had been found in mineable quantities. The usual rush morphed into the first use of float/ski plane

57

supported staking activity, and in 1927, Howey Bay was the busiest airport in the world! When the Howey Mine went into production at the foot of Howey Bay, they had up to 400 men to feed, and kept ten men in three camps in the bush, hunting moose to put fresh meat on the table. The point is this: Red Lake has always known how to get it done. No need to discuss a long list of requirements with the lumber yard. You just tell them you are building two 14x16 frame tents. You also call the Cash and Carry and tell them you need a four-man grocery order. They know you need salt and pepper, mustard and ketchup, and they know OCA will tell them when they need the order at the dock. These people have been at this for years - no problem.

Table scrap

One June evening I am heading back to Red Lake from Black Beaver. The aircraft is a Beaver, and the pilot is Eddy Cull, another one of the best of the best. We lift off at 9:30 p.m. in clear weather for the one hour trip, and ten minutes later, without a word spoken, Eddy pulls the pin on the control mast and swings the wheel to my side of the cockpit. Then he pulls a book out of his door pocket and settles back to enjoy a duster.

What the hell? I've got the controls, but it's not a dual control set up. I have no rudder pedals, and I waggle my feet and look at Eddy. He's still reading, but without looking at me he taps the artificial horizon and the airspeed indicator. Then he points to the rudder trim, a tiny wheel on the ceiling between us. I've flown this route in and out of Red Lake many times by now, but suddenly I'm no longer a passenger, and everything looks different – a lot different.

Did we cross this lake at the end of the bay? Why does that river seem unfamiliar?

I know you experienced pilots are already smugly smiling at the simple rookie, but I am processing a lot of new info and my brain is slow on the up-load. Why is my left wing low? I over-correct and now I'm turning right. Back on track, but now I'm losing airspeed – ahead a little on the control yoke. We pick up a bit of crosswind, and I'm dog tracking. My left hand is on the rudder trim and my left arm is getting mighty sore. I trim her a little bit left and rest my arm, but now I'm heading for Port Arthur, so I trim back and now I'm on my way to Winnipeg.

Eddy stirs, pulls out his map and lays it on my knee, and says "If you are going to fly in circles all night I'm never going to get to bed" and goes back to his book. By now the sun is disappearing in the west, a velvet curtain is drawing over the Northern Lakes, and some good soul in Red Lake turns on the street lights. I have a target! Ten minutes later Eddy puts away his book, takes the wheel and we skim in onto a glassy moonlit Howey Bay. What a magical evening!

(Ten years ago I was talking to a pilot at the seaplane dock in Kenora, and he told me Eddy was flying multi-engine at the Pickle Lake airstrip. Wherever you are, I hope you are enjoying yourself Eddy – good on you!)

So we fly in to Black Beaver and build the camp. The first order of business is to clear a few trees and build a good dock, and we do so. The first loads of lumber and stuff are piled on the shore, and by the time the next two Twin Beech loads come in, we have a decent dock ready for them. The rest of the day we lay the tent sills and start construction.

One more Beaver will arrive later on, and we will return to Red Lake for the night. Tomorrow the tents will be fit to live in.

The next day we fly in, and by supper time the camp is virtually complete with just some fine tuning in the kitchen tent to finish, and the boys can do that in the evenings. The two-way radio is operational but will be used sparingly. The boys will check in with OCA daily for messages, otherwise it is merely a safety measure. I set up a grocery schedule – they will be supplied weekly. We also pull out the grid maps and I assign work duties. Bob is experienced on the mag, Kenny and Tim will be on the JEM – I will catch a back haul to Red Lake. I have other things to do.

I head to Port Arthur and bring back the soil samplers. The load isn't as big this time, but we have another canoe on the roof and two more guys perched on sleeping bags in the back. I wish I had a Suburban to drive, or at least a bench seat in the front instead of the two buckets. Two canoes are all we need: one for the soil sampling camp and one for the geologists. Black Beaver doesn't need water transportation and the prospectors and line cutters have their own canoes. While I'm in Port Arthur, Don takes me down to the seaplane base and I meet John, the pilot of our leased Cessna. Yes – Noranda has a 180 on lease for a year, and John will arrive in Red Lake next week to fly me around in style. Pretty cool huh? Of course he is not exclusively at my beck and call but I will be using him more than 50% of the charter hours. The rest of the 180 time will be used to ferry bigger fish then me.

The next day I fly into Black Beaver with their grocery order and to make up the full load on the Beaver, we put on two barrels of av-gas. We will maintain a fuel dump at camp for our leased Cessna. I spend a couple of days to help Bob complete his base station readings, tune up some map work, and set up an organized system for ordering groceries.

Grub comes in every seven days, weather permitting, and when the order arrives next week's order goes back with the pilot. Dave at OCA will make sure it gets dropped off at the Cash and Carry. It can be a tad confusing if you don't remember what you ordered last week, and the pilot is busy. He can't hang around while you unpack the order and make up this week's order and another thing – you are supposed to be in the bush, right?

(Rule #39, sec D, sub-sec IV paragraph 7: If the pilot wants you at the dock, he will circle your work area while changing engine revolutions, and you hustle it on home. If the pilot doesn't need you, he drops the load on the dock, picks up outgoing mail and flies onward.)

So I have come up with a plan. I supply each camp with a handful of Jiffy Notes. Jiffy Notes are merely pads of lined paper slightly smaller than letter size. They come in three sheet units. You write your grub order on the top sheet and by some magical process the words appear on the next two sheets. You keep one copy and send the rest in. Now you have a record of what is coming in on the next flight and you can order another week's grub, confident that no unnecessary duplications will arrive. No one likes to waste food.

It's a great idea and Kenny agrees. We do up the next grocery order together and I institute my own employee benefits plan. Now, you must recall I learned my basic bush trade at Mother Inco's knee and she didn't allow any frills in camp – for example, bananas, and certainly no soda pop. I am more lenient. The pop thing I could understand in the early 60's, as canned soda drinks were not yet widely available, but this is 1969 and there is nothing more refreshing than a can of semi-cold carbonated beverage after a hot day on the picket line. Noranda has only one rule regarding grub. If you don't waste it you can get it, and I like

59

that. My problem is that I always have to give things one more push. At the bottom of the list I write "one carton of Cameo cigarettes (bury as meat)."

Kenny is the only smoker in camp, and he thinks this is pretty cool. We don't know that those three words at the end of the grub order will soon cause a seismic disturbance with its epicenter at an office high above Bay Street in downtown Toronto, with aftershocks radiating as far as my room in the Annex of the Red Lake Inn.

But the earthquake is a few weeks down the road – right now I have another mystery on my hands. On the morning of my third day at Black Beaver, I call OCA and Dave tells me I am to come back to Red Lake and call Port Arthur. That's all – no explanation – just get back here and call Don as soon as you get in. This is a real head scratcher.

I arrange a pick-up with Dave. I will spend the morning in the bush and will catch a back haul to Red Lake in the afternoon. I climb out at the OCA dock and say hello to Dave, but he seems to be a little embarrassed. I soon find out why. I check into the Annex and call Don. "What's up?"

"They've lost Chic and Leo," says Don, "You have to go in tomorrow and find them."

Well holy hell! No wonder Dave felt foolish. This was why the radio conversation was lacking in a few details. Dave didn't want the whole North Country party line to hear that OCA had lost a camp. I hustle back to the seaplane base to talk to Dave, and the mystery begins to unfold.

They had put one prospecting team, Chic and Leo, into the south shore of North Spirit Lake a little more than a week ago, and it took two Norseman trips to do so - Chic and Leo don't travel light. North Spirit is a big lake, at least 30 miles long if not more, but the OCA pilots know the lake and the guy that put them in is pretty sure of the camp location.

Yesterday another pilot had flown in with their grub order, but the camp was nowhere to be found. He cruised the south shore for almost an hour, but there was no sign of Chic or Leo. He returned to base, the groceries went into our storage bin and the perishables went back to the cooler at the Cash and Carry. I will go back to North Spirit tomorrow with the pilot who put Chic and Leo in, and we will take the groceries with us.

I go back to the Cash and Carry and tell the manager to have the meat and veggies at the dock tomorrow morning. I also give him the Black Beaver order for next week and point out the Cameo addition. He knows the drill, this is a common ruse in the game, and on the whole it is a pretty insignificant scam. I just shouldn't have written it down, that's all. As an afterthought, before I left the store I told him to throw in a carton of Belvedere cigarettes for Chic. (Man, these bush guys have weird tastes: Cameo and Belvedere – two of the most unsmokable of smokes.)

Sidebar: I once asked Kenny why he smoked Cameos. Did he like the menthol flavour? He said it was purely for economic reasons. Fellow smokers never bummed cigarettes from Kenny.

The weather is out the next morning but the clouds start to clear after lunch and at 3 pm we lift off Howey Bay in a Twin Beech with Jim Cousineau at the wheel and Chic and Leo's groceries in the back. This is my first trip in a Twin Beech and it is pretty cool. It's also the first time I have met Jim, and he is pretty cool also. Jim is another legend of the

North. (My goodness, I keep running into legends!) He is OCA's chief pilot at Red Lake. He is in his early 60's and will be retiring in a few short years, to own and operate the Gold Eagle Hotel and Bar on Mackenzie Island. He is also from Fort Frances, so we have a bit of a connection. He is a fine pilot and a really nice man. I like him immediately.

As we cross the Berens River we leave the scattered clouds behind, and before we reach North Spirit, Jim banks the Beech to the left. He tells me he has something to show me.

We are now 20 miles off course to the west, and Jim brings the Beech down to 300 feet. A huge fault runs south off the west end of North Spirit, and at the east side of the fault a 100-foot high vertical rock wall runs a mile or more down the fault line. It must be a mixture of quartz pegmatite with some silicas and mica thrown in, and in the late afternoon sun it shimmers and glimmers – it is a sight to behold. I am doubly amazed. The sheer beauty of the place is one thing, but the other thing that amazes me, is that this old pilot is willing to go out of his way to share this with a young doofus like me. Years later I still think back to that day. Jim had probably made the North Spirit run more than 1000 times, and the sight of that rock wall must have been a stress reliever for Jim many, many times.

We fly on down the south shore of North Spirit and Jim shows me where he dropped the two old geezers off. We land and taxi past the spot, and there is no sign of camp, not even a blaze or a cut-off tree branch. Jim can't understand it. He knows they were at this spot a week before. We take off and fly the shoreline again. On the second pass, I think I see something, and we land and taxi in. Sure enough, I had gotten a glimpse of their green canoe, upside down and almost invisible beneath the evergreens along the shore. As we get to shore, Chic and Leo emerge from the trees to meet us, all smiles and "How are you doing? I am sure glad to meet you," shaking my hand and trying to figure out why Jim and I are standing around open-mouthed and speechless.

Didn't they know we were looking for them? Oh man, are they ever apologetic. They thought the searching aircraft might have been stakers and they had not yet had time to stake a claim block on their showing. (Prospector paranoia – if you don't have your showing protected with claims, you don't advertise your presence.) They had spent four days building camp and had just finished staking their claims. Tomorrow they will build a dock and they are so sorry they have caused people to worry. Jim and I say "no problem" and the teapot is put on the stove. As we wait for the teapot to boil we haul the groceries up and I get a chance to check out their camp, and what a camp it is. These boys have honed the art of building a home away from home to perfection. The tent is a huge 18x20 heavy canvas jobby and in front an extra fly makes a comfortable patio/outdoor kitchen. It's a standard fly camp set up, but their living area is completely free of small bushes, twigs and other debris found on any forest floor. Inside the main tent are two comfortable bunks, a map table, a sample table, and a shelf holding a small library of books. I'm not kidding! And we do not "just pull up a stump," no sirree bob! These stumps have backs and arm rests! And I see why they were so hard to spot – the tent is forest green. Years of prospecting paranoia are evident in their choice of campsites. They always build amongst high evergreens and they take care to never cut down a tree near camp or shore for their camp construction. They are invisible without heat seeking imagery.

They are still very apologetic about moving camp. What happened was that they had checked their maps last week before starting to build. They realized they were in the wrong place, so they had moved the whole shebang four miles east. Mystery solved!

Before Jim and I leave, Chic finds the carton of Belvederes and wants to pay me for them. "No problem" I tell him. "These are on Noranda." Chic is almost speechless.

"Twenty-two years," he says, "Twenty-two years it takes to get a free smoke!" As the Beech pulls out of North Spirit everyone is happy. It's been a good day after all.

Sidebar: Further to Chic and Leo's traveling circus: Around 1960 they were prospecting between Atikokan and Fort Frances. The Hwy 17 link between the two towns was not yet built. CNR ran a local mixed freight and passenger train - it would stop anywhere. Chic and Leo flagged it down, and the train crew helped them haul the Taj Mahal up the side of the roadbed to the baggage car. Where were they going? Five miles on to the next lake. Total transportation fee? Five dollars each.

Sidebar: When the original railway charters were drawn up in the 1800s they included a clause that any prospector's canoe was to be transported at a flat rate of one dollar. Two, two hundred, or two thousand miles – one dollar! Try it sometime.

Table Scrap

Another day later on I am in the right-hand seat of a Beaver and once again Jim Cousineau is the pilot and once again I get a side tour. We are a fair piece east of Red Lake – why, I don't recall. We were probably moving a fly camp.

The new Uchi Lake mine site is not far off our flight line so Jim gives me a birds' eye view of the hustle and bustle. Then he asks me if I have ever seen the old Confederation Hotel. I haven't, so we fly on another mile or two or ten.

The mine had closed years ago and there is no trace of the townsite, but the old hotel stands two stories tall and strong. It is boomerang shaped with one wing to the northeast and another to the southeast. The central section is a bit taller and a large fieldstone fireplace and chimney is flanked by windows and verandahs. Jim tells me it was quite a place in its heyday and has so far has escaped vandalism.

There is evidence of a once broad, half-circle driveway in front of the hotel. Here sits a '25 to '27 model T four door Tallboy – all there, all intact except for the fabric roof insert. This was the hotel jitney, used to ferry guests to and from the lake in all-weather comfort.

We do three low-level passes. I'd like to land and do a walk-about but there is work to be done. As we bank off to the northwest I once again thank Jim for a tourist interlude.

Sub-Scrap

Four years later I saw the "T" again on Hwy 105 while returning from a propane delivery to Red Lake. South of Perrault Falls I rounded the corner and there she was, sitting high and proud on a two-axle trailer. Someone had taken advantage of the new Ear Falls/Uchi Lake road to bring her out and I assumed they had used winter ice and frozen swamps to extract her from Confederation.

This time I had a closer look at her. The original black paint was now a brownish colour – a proud patina befitting her age – and no rust holes – and not one fender dent. She was straight and true and still had all her window glass. She also sat level – wire wheels instead of wooden spokes confirmed her 1925 – 27 manufacture date.

At Vermilion Bay she turned left and I turned right. We were both going home.

The next day CF-KHM, arrives in Red Lake with John flying. I meet him at the OCA dock and check him into his summer home at the Annex. Before he lands I have a chat with Dave. I feel a little guilty about having our own plane at their dock, but Dave tells me not to fret about it. They are charging us dock space and selling us av-gas and besides, OCA has enough trouble keeping up with their flying demands. Our airplane will only be doing some grub hauls and taxi service. Most camp moves are still in Dave's hands.

KHM is a snazzy, freshly-painted blue and white Cessna 180, with a zero-time time engine when it left Port Arthur, and John is a high hour pilot and a nice guy. It should have been a union blessed by the gods, but I will soon find out that the relationship is mildly dysfunctional. John doesn't like KHM, and KHM doesn't like John. I will end up being their marriage counsellor for the rest of the summer.

Sidebar: John is not a bush pilot, neither figuratively nor literally. Almost all his previous hours were as a flying instructor – you would think this would be a plus. Maybe wheels on pavement are more forgiving than floats on water and John freely admits that he doesn't have much float time. Also, he seems obsessed with the instrument panel, tapping this, and adjusting that. All pilots check the gauges. They just don't make a big deal about it. John was a flying instructor – therefore, maybe he was in the habit of stressing the importance of the instruments to his students.

Restored Fire Tower at Temagami, Ont.

Table Scrap

Ray Cameron, one of my high school buddies, spent many years flying hunters and fisherman in and out of tourist camps and lakes in the Rainy Lake area. Sometimes he would get a real jerk in the right hand seat – one of those guys who had been there done that and had the t-shirt. Ray would get tired of the BS, so he would suddenly lean forward and tap the oil pressure gauge, adjust the throttle, and tap the gauge again. This always shut the guy up and he would keep his eyes glued to that gauge. Kind of mean, but very effective.

Sub-Scrap

In '55 or '56 when Ray was seventeen or eighteen he spent a summer manning a fire tower in Quetico Park. It was a one-man operation. He had a radio in the tower but it was for official use only.

He had the occasional visit with canoers paddling by, and sometimes they would climb the tower to check out the deal. He told me he would keep them talking as long as possible – poor buggers.

I saw him that fall after he came out and I asked him what he had learned. He said he knew exactly how many peas were in a can, and how many beans in pork and beans. The pork was easy to count – always <u>one</u> piece of pork.

But here's the best part. One day he had a visitor in the tower when the plane brought in his grub order. As they watched the pilot unload at the dock the visitor asked Ray what he wanted to do further on in life.

Ray pointed to the pilot. "I want to be him – he goes home every night."

And the rest is history.

I learn to work within John's capabilities. He doesn't like heavy loads and that's fine with me. He wants a full mile of lake runway – fine with me. He doesn't like gassing up his own plane. Ok, there is always a young guy at Black Beaver willing to do the fuel transfer, and at OCA the dock boys do the job.

The first order of business for John and I is to fly two grub orders into two camps. One order is for George and his partner (the other prospector duo) and one is for the student soil sampling crew. I want to catch the students in camp. I have some maps for them and will bring out samples to take to Port Arthur and I also want to check up on these young fellows.

We leave Red Lake at around 3 p.m. and I am a bit perplexed. KHM takes forever to get off the water, and the airspeed indicator shows almost 60 knots before John pulls her off. Finally we are in the air and on our way.

As we land at the sampling camp I see their canoe tied up on the east shore of the lake, so the kids must be still in the bush. That's OK, we will unload at their camp and wait for them. They will hear us come in and will return to camp, which is on a nice pleasant point on the west side of the bay. As we taxi up to the dock the three boys come down to meet us, and they look kind of glum. This is strange: three guys in camp and the canoe across the lake? What's going on? It doesn't take long to find out. Someone has stolen their canoe! They had tied it up at the end of the picket line when they went in that morning and when they came out for lunch the canoe was gone. I know what has happened – they had come out on another line and had forgotten which one they had left the canoe on. John opens his mouth to say something and I give him the high sign – we are going to play this a bit. "Hmm, who would steal a canoe? Have you seen anyone else on the lake?"

"No, but it's a pretty big lake. Someone must have passed, saw the canoe and lifted it."

"This is a big problem, John and I will have to report the theft and find you another canoe, and we need something in writing so the police can start their investigation. How about the camp? Obviously someone will have to stand guard from now on. Do you want a gun?"

Finally we can't stand it any longer. John and I fess up and we all laugh like crazy. John taxis one kid across the bay to rescue the cedar strip and I take care of business. Give those boys 'A' for effort. They had walked 6 miles around the swampy end of that bay to get back to camp. They never lost their canoe again that summer.

Before we leave camp I notice that they have a 90-degree elbow on top of their stovepipe where it protrudes above the tent fly. They tell me this is to help the wind draw smoke out of the airtight stove. West wind – face the elbow east – easy breezy. Now I give them two more 'A's – one for initiative and one for inventiveness. I don't tell them that some night a weather front will pass through, the wind will shift to the east, their tent will fill with wood smoke, and they will be pulling that elbow off in a 2 a.m. rainstorm. Some things you just have to learn yourself.

Sidebar: A couple of weeks later I join John on another grub run, and the boys are sure glad to see us. The 20-lb propane tank on their jiffy range ran out 4 days ago and they can't change to the fresh tank – the threads are seized. I pick up their 12" adjustable wrench and in 2 minutes I have the tanks switched as they watch with gaping mouths. They didn't know that all propane tank connections have reverse threads. When they tried to back the connection off, they were actually tightening it. This was really my fault. When I put those boys in to that camp, I assumed everyone knew about propane. I learned a lesson myself that day "Always assume the worst and always hope for the best."

The students didn't starve though, they had built a dandy stone fire ring and cooked their meat on a cast iron skillet while the potatoes boiled in a pot hung from a stick on the other side of the fire. I've always hope that those three guys went back to school that fall, proud of their summer experience. They did their job well, always cheerful, and with not a complaint all summer long. I loved those kids.

On to George Cates' camp, and once again I am amazed (I seem to be amazed a lot this year.) In the broad spectrum of camps I have built, lived in, and visited, I would put Chic and Leo's at the ultra violet end, with George's camp in the infrared range. Chic and Leo like dark, cosy and seclusion from prying eyes. George likes white, airy and in your face. Both camps are so clean you feel you should take your boots off before entering, but George's positively glistens.

He is on a nice flat outcrop shore, and hardly a tree or bush interrupts the breezes. Some shrubbery is round about, but it's almost like an English garden – as far as I know, George might trim the hedges. Inside the tent, George has started an interior wall around the inside of the sail silk. Every night he adds little spruce poles, trimmed and partially squared to 2x2 size, and builds a miniature cabin wall. My boss, Don Cross, tells me that if you leave George in one spot over the summer, he will have the tent completely finished inside.

Sidebar: George would always call me "Don" and then apologize. I told him it was OK – I probably looked like a Don to him. In the mid '70s I stopped to chat with George at his retirement home in Kingsford Twp. north of Emo. George was building his log cabin again, a real one this time, with doors and windows.

"One round a day Don, "says George. "I do one round a day, rain or shine."

If that house ever had to be moved, I can imagine the movers screwing a huge eye bolt into the ridge pole and lifting the house with a crane – it would have been that solid.

I have to go back to Black Beaver for a few days. The boys are not the best map-makers and I have to tune things up once in a while, and I also have some maps for the line cutters. The weather is out, so John and I spent a day twiddling our fingers. The next day things don't look too bad by lunch time, so we give it a shot, but at the Berens River we are confronted by a wall of clouds down to the tree tops and we turn back. The Berens runs almost straight east/west where we cross, and many times that summer the phenomena (in my mind) would occur. The Berens seems to be a weather divider – good weather on one side, bad on the other. This must be caused be a slight difference in elevation along the river I guess. No one frets it though. You either go on or go back – no problem.

The next day we take off into a strong north headwind, and for a change KHM is eager to get off the water. We have groceries for Black Beaver and the cutters, and on the way in I show John where the line cutters are camped. They are on a small lake off the east end of Black Beaver and are almost finished the east end of the huge grid. I have maps for their next line cutting targets.

Their lake, although not large, runs north/south and is at least ¾ miles long, adequate length for a 180, and certainly doable with a strong north wind, but John doesn't like it at all, at all. I'm starting to get a bit pissed with this guy. I say, "Ok, we'll drop Black Beaver's groceries off and discuss the situation."

So we bank left to Black Beaver and land in a strong crosswind. Now, this lake is six miles long and almost a mile wide. There is a bit of ridge with tall spruce trees along the south shore, but any competent bush pilot could drop her in crossways with room to spare considering the strength of the north wind, but not John.

It's a bit of a white-knuckle crosswind landing, and just before KHM drops off the step, a strong gust of wind hits the starboard wing and the port wing goes under water! No kidding! The 180 lurches to the left, John pulls the wing up and we taxi to the dock in silence. The boys are at work and the camp is empty, so I put on the coffee pot, carry the groceries up to the kitchen and John and I have a talk. I am going to break my own rule – I am going to argue with a pilot.

It's not really an argument. I merely point out that I have to see the line cutters, and although I could walk the 5-mile baseline to their camp, I can't pack their grub. We have two alternatives. Either John puts me in there, or we return to Red lake and I return in an OCA Beaver. The words "with a real pilot" are unspoken but tacitly understood. We reach a compromise. John is light on fuel, so he will drop me in, return to Black Beaver and (heaven forbid) refuel KHM himself, and return to town. I will walk back to camp that evening. I am going to spend a few days there anyway, and I can catch a back haul to Red Lake later to get

67

home. Good enough – John drops me and the grub off at the line cutters' camp, and we watch him pull out of there empty and with plenty of room to spare.

Sidebar: I never did rat John out on the wing dipping, but the incident did cause a subtle change in our relationship. I still got along well with John, but I was a little more circumspect in flying with him. I tended to pick my spots from then on, preferring to use OCA whenever possible. John knew where all the camps were by now, and I would be more often a director and less often a passenger. He could unload his own plane – time to grow into the job, my son.

At summer's end, with the camps coming out, I will learn that John had a habit of washing his outboard navigation lights in lake water. Some of the boys had started to call him "dipsy-doodle" behind his back. I felt kind of bad about that, John always tried his best, and that's all you can ask of anyone.

The head line cutter is Roland Sevigny from Val d'Or, P.Q. and we click right off the bat. I give him the map for his next two jobs and we arrange to move him in three days. We have a cup of tea, and I guess he is sizing me up. He has something to show me, and wants to make sure I am not a blabbermouth. He says "Come with me," and we climb into his canoe and head north across the lake.

Black Beaver deserves its name – its water is not all that dark, but the lake bottom is made up of black silt and it seems to swallow light. Also, as in many lakes of this type, Black Beaver is populated sparsely with small northern pike. Roland's Lake is the opposite – crystal clear and cold, spring fed no doubt. At the north end a narrow creek travels through a small valley to another little lake, not much more than a pot-hole.

As we enter the creek Roland lifts the three-horse kicker, we pick up the paddles and I see why Roland doesn't want to use the motor. The creek bed is filled with fish! It's like Main Street in Walleye City on Saturday night, with fish swimming north and fish swimming south, and none of them seem to be worried about our canoe floating above them.

Roland says they come out every afternoon after work and scoop up their supper with a landing net. We don't take any fish today. The line cutters have fresh meat for supper.

We return to the camp, I eat with the crew and walk the five miles back to our camp in the evening. My bedroll awaits.

(I will run into Roland often in the next few years. In fact, he will be one of the men sitting in that bathtub in Savant Lake in '71. I liked Roland.)

Sidebar: You will recall my theory in Book One about seagulls stocking lakes? On the way into Black Beaver one day later that summer I had John fly me over those two lakes again. There was no sign of a creek feeding the larger lake, nor was there a clear waterway exiting the pothole, just a large swamp running on north to North Spirit, about five miles away. It had to be seagulls or pelicans who transplanted those yellow pickerel.

Ten years later I tell this story to a dedicated sports fisherman, and he tells me that there are many lakes like that in the Shield, too small to entice a fly-in group, and too small to support a large fish population. He says that those fish would have had large heads and small bodies – harvesting these potholes would make for healthier fish.

A few days later I catch a back haul to Red Lake and check into the annex. The desk clerk tells me I have an urgent message from Port Arthur. I am to "Call Don AT ONCE!"

Now, when a message ends with AT ONCE, nothing good is going to be in store, so I first have a nice hot shower while I go over a list of transgressions I may have committed. Finally I make the call, and Don is apoplectic.

"What in the hell are you doing up there?" he yelps.

I am mystified – I'm working, that's what I'm doing. Why has he got a knot in his tether chain? "What's the problem?"

"I'll tell you what the problem is." says Don, "The problem is that Ralph has a copy of a grub order on his desk in your handwriting, with a carton of cigarettes on there, noted "bury as meat," and I have been taking flak from him for two days, and Ralph wants to know what kind of crooks do I hire, and he wants answers!"

By the end of this tirade I am holding the phone about a foot from my ear to save my eardrum.

Well. I'll be darned – that dumb Cash and Carry had sent a copy of that grub order in with the invoice. How stupid could they be to do that?

I explain to Don what happened, pointing out that it was just one five dollar carton of smokes, for goodness sakes, and I also point out that they were not for me – I've been off the cigarettes for more than a year.

Don is somewhat mollified, and he says that he will calm Ralph down. He also indicates that he is not averse to a bonus now and then, but he wants it understood that he knows nothing about it – it's up to me.

I don't mention Chick's Belvederes, and I stop supplying smokes – It's not worth the hassle. However, I do tell the Cash and Carry manager of his faux pas. He is very apologetic, and of course, blames the office girl. All is forgiven – they have been as busy as one-armed paper hangers this summer. Crap happens.

Less than two weeks later, I get another early morning "at once" message. Don wants me to bring the spare canoe to Port Arthur "at once," so I tie the canoe on the GMC racks, throw in some life jackets and paddles, and drive eight hours to Port Arthur.

The next morning I hit the office, to find out what the emergency is, and Don tells me Ralph is coming up from Toronto with his two sons. They want to do some paddling.

Well, Dirty Dinah! A bonus carton of smokes is a federal offense, but a two-day return trip to haul a canoe for Ralph's personal use is okay?

I head back to Red Lake before Ralph gets to the office. I know I won't be able to keep my mouth shut.

You know how the saying goes - corporations are like septic tanks - the big turds always rise to the top. Maybe it's my bad habit of always identifying with the working man - but I will lock horns with Ralph a couple more times before I leave Noranda.

I head back to Red Lake with a few things on my mind. Don has told me that a graduate geophysicist has been hired, and will join him at Port Arthur to coordinate geophysical activities in North-Western Ontario. I ask him where that leaves me, and Don says not to worry, I will still be geophysical foreman. As I drive north I cogitate on this new info - it seems to me we are getting a little top heavy here, but as I pull into Red Lake I put my worries aside.

The next couple of weeks are uneventful. We are like a well-oiled machine, camps are moved, grub goes in, I help out with the map work at Black Beaver, and on June 21st I head home for a break. I've been calling my wife every three days or so, but I miss my family and I stretch my stay out as long as I can. Five days later I am on my way back to Red Lake.

I receive my smooches and hugs before I leave and my three-year-old has tears as usual, so she gets an extra hug. Whoever said parting is such sweet sorrow sure nailed it but I'm thinking it's a little heavier on the sorrow side, as far as I'm concerned.

I stop in Emo and pick up two new hires, both recent high school graduates, and this is one part of my job I like. I can usually hire who I want, when I want.

I always like to get on the road early and then stop for a bite, so we pull in at a restaurant in Nestor Falls for toast and coffee. At a nearby booth, three young fellows sit, laughing and joking and my new guys tell me they are from Fort Frances, and are working for Inco.

"Well they're not working very hard," I say. It's 9 am on a clear day, and they are rehashing last night's party. Five minutes later in stomps a guy who I had worked with in 1966 in Smooth Rock Falls. I have heard he is an area supervisor now, and here he is chewing out his men, screaming at them actually, and then storming out with only a curt nod for me. I ask them if they don't like their job, and they laugh. "He can't fire anybody" they say, and I feel sorry for the guy. That's Inco – they give you a ton of responsibility and cut off your testicles. Tough row to hoe for middle management, I'm thinking. My two new men agree.

Rod and Lorne are the two new hires. I had worked with Lorne's dad in the paper mill and we became good friends. I figured if Lorne had any of his dad's work ethics he would be ok, but that will prove to be wishful thinking. Lorne is quite willing to do what is asked, but the drive and initiative required in this line of work is just not there. He will stick it out until next spring and then move on to something more suitable to his disposition.

Rod Swire is another kettle of fish; happy and outgoing, and built like a brick outhouse. He is going to be ok, I figure, and I figure right.

As we head up to Vermilion Bay and turn off on Hwy 105 to Red Lake, we chat about the job. I, of course, am telling stories in which I am the hero.

A couple of days ago I had visited my barber for a haircut, and he is another old friend. His wife has just produced another son, and as I leave the barber shop, Ron gives me a cigar. I don't want to throw it away, and I don't have anyone to pass it on to, so I had tossed it on the dash of the G.M.C and it had been sitting there for three or four days, whispering "Smoke me, smoke me!" Ten miles up 105 I give in. I find a match, unwrap that sucker and light up, and boy oh boy, does it ever taste good! I stop in Ear Falls and buy a pack of Player's Filter cigarettes, and 500 plus smoke-free days go down the tubes. I've smoked ever since.

In Red lake we meet Peter, the new geophysicist, who had flown in from Port Arthur with John that morning. Peter and I go to the bar for a chat. I turn Rod and Lorne loose to check out the town - tomorrow they go to the bush. Peter tells me he will not be doing my job. Uh huh – we'll see.

Early next morning we fly in to Black Beaver. Rod will train on the mag, Lorne on the JEM. Peter is along to observe and learn – he says he just wants to get his feet wet. Today I will try to make his wish come true.

I show Peter the camp and the maps, and I tell Bob and Ken to break Lorne in. I will take Rod out with the mag. I give Rod a short tutorial on mag operation and dress code. Rod has a metal belt buckle and a spiffy new hunting knife. I tell him to turn the buckle to the back and leave the knife in camp. He also has a little $4.95 compass. I laugh and show him my fluid filled Silva Prospector model. "THIS is a compass", I say, "Yours should have a picture of Mickey Mouse on it."

We will be working on a cut and chained grid and a compass is unnecessary. Rob wants to carry his compass though, so I tell him to keep it in his hip pocket if he is reading mag. Today I will be doing most of the reading.

The west end of the grid has not been covered yet and just west of camp the beavers have been working to enlarge a pond. I have seen it many times from the air and Roland was unable to cut two lines due to open water. I want to pick up the line running down the west side of the pond, cross over on the south tie line and read a line on the east side of the pond, ending up back on the baseline. Now I will have the water bracketed. We plan to return next winter to fill in the gaps.

So, Rod, Peter and I head out and we are a strange looking trio. I carry the mag, Rod carries the notebook and Peter has his hip waders on. Hip waders – no kidding! Peter has brought a pair of chest high waders with him and he is wearing them. Now I've seen everything! It is a hot, hot, muggy day with a low over cast sky, and not a breath of air is moving. I am already sweating, and I figure Peter must be having a steam bath in those waders.

We pick up our line, do a base line check and head south. I keep up a running dialogue with Rod as he writes down the mag readings and we reach the south tie line and take a break. Then we walk 1200 feet east on the tie line, pick up the line we want, and start to read back north. When we hit the base line we will have read 3 miles of mag – not a bad day considering I am training a man.

Within 500 feet we run into water. I know what has happened, the beavers have been raising the water level in their dam. The terrain is flatter on this side of the pond and the pond is much larger than it appears from the air. Heavy bush has made it hard to spot water from prying eyes in the sky, so I study on things for a bit. Do we skirt the water? I decide to forge ahead, surely the terrain will rise. Soon Rod and I are in over our knees. Peter, of course, is dry as a bone in his hip waders and is looking pretty smug, and he's starting to piss me off.

The water gets deeper, and I'm not pretending to take readings any more. I'm wet to the ass and Rod is up to his belly button, and we are holding the mag and notebook above water. I'm looking for holes in the forest floor – I just want to get college boy wet.

Finally I give up and we head east to find high ground. When we emerge from the depths and pick up a line, we are another 800 feet east of where I expected to be. These beavers are an industrious bunch.

We find a log, Rod and I take off our boots and wring out our socks, and I tell Peter he's a smart ass. "I know you were trying to sink me," he says, and we laugh like crazy.

It's too late now to try to read another line back to camp. We've been playing in the water for two hours. "Let's go home," and I start off on the picket line. Rod has Mickey Mouse out "I think north is that way," says Rod, pointing south (I think.)

"That's the trouble with those contraptions," says I. "They can't be trusted,"

"Follow the old prospector," says the old prospector, "We never get lost".

A hundred feet further on I check the next picket, and holy hell, I'm going south! Mickey Mouse is right on! All that traipsing around trying to get Peter wet has turned me 180 degrees.

Without a word I head north for the mile walk back to camp and on the way I hear sounds behind me. It's either squirrels, or someone is having a bit of a chuckle.

We hit camp and I strip off my wet clothes, run down to the dock and jump into the lake to cool off. As I dive into the water I feel a sharp pain in my groin – what's this? Did I hit a branch or something? I paddle back and lift myself onto the edge of the dock. My left testicle is the size and color of a grapefruit and the pain is sharp and intense. I can't see any outward damage and as I stand up the pain goes away and everything returns to normal. I get into dry clothes and tell Peter I am going back to Red Lake with him. I want to see a doctor to solve a mystery.

We have supper in camp. Summer solstice has just passed and the evenings are long. We have a two hour wait for the Beaver back haul and I spend the time going over the JEM and mag maps with Peter.

After supper Rod pays a visit to the little plywood shack outback. He returns looking a little worried and asks for a wire coat hanger. Apparently he had put his hunting knife on and when he undid his belt in the biffy the knife had slipped off and had fallen in the hole. Rod wants to retrieve it. "It's only a knife," I tell him, "Buy yourself another."

"That knife is a present from my dad," says Rod, "I'm not going to leave it in a shit hole."

It takes Rod almost an hour to dig that knife out and wash it off in the lake. Right then and there, I know this guy will last and I am right on.

Sidebar: In 1971 Rod would work for me for a winter season. He then returned to Noranda and became their geophysical foreman (my old job.) Rod stayed with Noranda until they disbanded the Thunder Bay office, then went on his own doing contract staking and geophysics. Rod remains a fond memory and I am proud to have started him on his career. That streak of stubbornness Rod showed me at Black Beaver is one of the prerequisites of a natural born bushwhacker. Good on you, Rod.

On the way back to Red Lake, Peter gets airsick. I am riding shotgun and I never get airsick, but it's pretty tough in a poorly ventilated Beaver, on a hot summer evening with someone in the back throwing up in his boot. I am prepared. I always carry a chocolate bar when I fly with a rookie just in case I have to calm down my own tummy, so I slip the O Henry out of my shirt pocket and unwrap it. He's honking and I am munching, and the sight of me munching makes Peter honk all the worse. For months afterwards, Peter tells everyone the story, but it's no big deal to me. As he washes his boot out at the OCA dock I toss my candy bar wrapper in a garbage can and say, "I got you, Mr. Hip Wader Smarty Pants," and we laugh like crazy.

72

The next morning I see a doctor and he explains to me what happened at the Black Beaver dock. It seems I am lacking a bit of important stuff in the cojone area. Mother Nature, since time began, has supplied some restrictions in the ball department. In hot weather, things hang lower to prevent overheating, in cold weather Mother Nature pulls the string and things tucks in a little closer. When I jumped into the cold water yesterday the control cord snapped back, my testicle twisted and cut off the blood return system, causing the swelling and pain. Most men have tissue surrounding the testicle to prevent this twisting, but I don't. There is good news and bad. The good news is that my nut untwisted right away. The bad news is that if it hadn't untwisted I had about 4 hours to get to a hospital, otherwise gangrene could result in my losing half of some very important parts.

Well that does it. For the rest of the summer I never again stay overnight in Black Beaver. My room in the Annex is now booked solid, and although I spend just as much time in the bush as before, I watch the weather and my plane lifeline very closely: Very Closely!

Don pays a visit to Red Lake a few days later, and I tell him about the situation. Here's a handy hint. If you want something from the boss bring your cojones into the conversation. He will start to fidget, surreptitiously adjust his own equipment, avoid eye contact, and agree with everything – probably a good time to ask for a raise while you're at it.

The doctor tells me that a simple operation will give me some scar tissue to make up for Mother Nature's oversight. Simple to him, maybe, but I'm doing some squirming myself as he discusses this so-called simple operation.

The summer goes on and the work gets done. Camps are moved, men get shuffled, and I fly to Port Arthur a couple of times with maps and reports. To tell the truth, I don't like the office scene much. Nine to five has always been boring to me, and flying in and out doesn't give me a chance to swing by to spend a day or two at home.

The good part about spending every night in Red Lake is that I am able to expand my social networking skills in the Bucket of Blood. The bad part is that my budget is taking a beating. I've been spending most of my rainy day afternoons and evenings with Sam Duggan, Noranda's contact diamond driller. Sam lives in Red Lake. His drillers are working in the area, and Sam likes his rum and coke.

Sam is an interesting character, and the more I get to know him, the better I like him. He started his mining career in Geraldton in 1938, gravitated to Red Lake after WWII and has been a millionaire at least once, and maybe twice, I am told. He is loaded with interesting stories, and what Sam doesn't know about diamond drilling isn't worth knowing.

Sidebar: Sam had enlisted in WWII with the intention of flying. He began training at the Portage La Prairie Commonwealth Airbase, and you might say that he had some landing issues.

One day his CO called him in and suggested that Sam would be better suited as an airframe tech. If he kept flying they would soon run out of planes.

Sam did get his steering ticket later on and would fly his own 180 or two for a number of years.

But Sam is well liked and respected in mining circles, and over the years he has made a ton of friends and a few enemies. One afternoon in late August I meet one of the latter variety.

I had spent the morning at Black Beaver closing the camp. The tents will stay put until I return in January, so we add some bracing and extra poles under the flys to handle the snow load. All incidental camp gear is stored in the tent. The two men left in camp will be dropped off at another camp with the left-over perishables from Black Beaver. Non-perishable items and kitchen stuff are left behind for the winter season. We secure the camp, drop off the men and groceries, and return to Red Lake at 3 p.m. I decide to check out the Bucket. Maybe I'll have a beer before supper.

The bar is almost empty. Sam is at a corner table with two guys and he calls me over and tells me to take a load off – he's buying. Three beers later I begin to realize something nasty is going on. Sam is pretty much hammered, but the other two aren't. As I watch and listen I notice they don't drink much, but they make sure Sam is getting more than his share. I'll call the two guys Tom and Dick. I've seen Tom around town many times and I don't like him much. Dick I know only by reputation. He owns some equipment contracted to the ongoing sewer and water work, and is reputed to be one tough hombre.

I see what is going on. Tom is pissed at Sam for some reason, and he is digging and digging at him, trying to get Sam mad enough to fight. Dick is Tom's hired gun, and it looks like Sam is in for a licking. Less than a half hour later, the crap hits the fan. Sam is really drunk by now, and he jumps up, and so does Dick. Tom sits there with a smug look on his kisser and he is keeping an eye on me. I don't worry about Tom. He is a fat slob and I have no fear of him. Dick is the one I've been watching.

As Dick jumps up to nail Sam in the nose, I step in behind him and put a hammer lock on him with my left arm. What the hell am I doing? I'm no fighter! I've never been in a tussle since grade eight – but I hang on. I've got 6 inches on Dick, and I'm in damned good shape and he can't break my hold. "You'd better let go of him Bob," says Tom.

"I can't," says I. "If I let go he'll hit Sam."

"You have to let go" says Tom, "He's turning blue."

I let go, Dick sits down rubbing his neck a bit, and they soon leave quietly. Sam is as loose as a goose. I snaffle his car keys, drive him home, and walk back to the Annex.

I play it cool for a few days wondering if I'm on the hit list, but nothing happens, and I put it out of my mind. Gone and forgotten I figure. Wrong again, as I will find out six years later.

August is almost done and so is the Red Lake work. Don calls me to tell me for crying out loud get your operation over with, we can't keep you in a hotel room forever. Fair enough – I head home leaving Peter to tie up any loose ends. I am going to stay home for a month. My summer is over.

Table Scrap

In the mid 70's I am living in Vermilion Bay and working on the pipeline. One Saturday afternoon I drop in at the local bar and by golly there sits Dave, the OCA base manager. We are glad to see each other and swap a few stories about what's new and what's not, and Dave drops a bombshell on me. "You're not the fastest gun in town anymore" says Dave.

What does he mean, "fastest gun!?" I am totally mystified and I ask Dave to elaborate.

It seems that my defense of Sam had reached legendary proportions around Red Lake. Dave says that Dick became a changed man, no longer belligerent and feisty – I had tamed the tiger. Three brothers have recently moved into town to work in the mines and they are now cleaning house in the bar every Saturday night. So six years later, I have lost my title by default. I laugh like crazy, but I'm thinking how lucky can I be! Had I dropped in at Red Lake any time in the intervening years I might have had to slap leather at high noon.

Whew! I figuratively wipe the beads of sweat from my brow.

Table Scrap

Fast forward to 1996, and I am sitting with Sam in his living room in Fort Frances. Sam has only five months left before lung cancer takes him away, and for three weeks I have been staying with him every night, rising at 5 a.m. to open our store. Stories about the old days are coming too fast to remember, but this evening Sam tells me what caused the showdown at the Bucket with Tom and Dick. Why that old fart! Here I thought he had been out of it that afternoon so long ago, and now I find out he was playing possum.

Suddenly some things are cleared up. For the next few years following the Bucket incident I will get a few breaks, and now I realize that Sam was behind some of them. Sam never forgot a favor, he just didn't make a big deal of it, that's all.

Table Scrap

In January of 1997 I spend a couple of days with a friend touring my old stomping grounds. He has contract snowplows at Sioux lookout, Ear Falls and Red Lake, and is checking out his equipment. We wrap up at Red Lake at noon and I suggest a beer at the Red Lake Inn. We walk into the Bucket of Blood and its déjà vu all over again. There is Billy at the bar, and behind the bar is Ralph, the owner. Agnes is at a table waiting for the next of a long, long line of lonesome high rollers. It's 1969, and I am once again 30 years old, and Ross and I close the place up. It's my last Bucket party, and it's a dandy!

Table Scrap: Jack Edwards and his Fox Moth.

Jack has a construction company in Kenora with a diamond drill division, and Jack flies a vintage Fox Moth, a later version of the WWI Tiger Moth. It is a fabric covered biplane with the pilot sitting in the open air a la Snoopy. Behind the engine a door opens into the fuselage and the plane can haul two passengers or cargo therein. The only drawback in my mind being that the passengers can't see where they are going. Their only view is through side windows.

One day I'm at the OCA dock waiting for some men to come in, and Jack lands his Fox Moth on Howey Bay and taxis towards the dock. As he nears the dock I start to get a little bit nervous. The only empty space is between a Twin Beech and a Beaver, and it's not large enough to park my GMC! I check around me - dock boys are loading and unloading planes, and no one is concerned but me.

Jack chugs in downwind, and at the last moment he kicks the tail around and slides right in, slick as a breeze. I tie off the float ropes to the dock and Jack climbs out nodding his thanks. My real reward is my view of his wife's ample cleavage as I help her out of the passenger door - thank you Mrs. Edwards.

Fox Moth on Floats

CF-DJB

Table Scrap

Another day I arrive at the dock to find a Canso tied up and dock boys are on top of the wing filling the fuel tanks. The Canso is the Canadian version of the PBY, and the PBY was used extensively in WWII, for search and rescue and spotting submarines. OCA flies this one out of Kenora.

Consolidated PBY-16 Canso

I am told that the Canso with full tanks can remain airborne for 26 hours. What the dock boys are doing is filling the inboard tanks with fuel oil. The outboard tanks have sufficient capacity to fly to their destination and back - the fuel oil will be drained into storage tanks up north. Pretty cool huh?

Table Scrap

I never needed an alarm clock in Red Lake. My room in the Annex faced Howey Bay and there is nothing that will get your attention at 5 a.m. like an aircraft starting a takeoff run outside your window.

One day I told a couple of pilots they were cutting into my beauty sleep, and the next morning, every dam plane that left the OCA base made sure they taxied down to the Annex before swinging around and cracking the throttle. I sure straightened out those boys, didn't I?

High Grading (Hi-grading)

Hi-grading is a common term for the practice of helping yourself to a share of the gold you are mining.

In the thirties, for instance, even though the hard-rock miners were paid relatively high wages, ($8 a day or more if you were in a bonus stope) the lure of a bit of hi-grading was hard to pass up. Gold prices were regulated in the $18 - $23/oz range at the time, rising to $35 during WWII. Every miner knew someone who knew someone else who knew how to funnel gold out of Canada to a country where someone would pay a premium price for the commodity. Good old free enterprise!

It was a risky business. It wasn't your gold to start with, and besides, until the price of gold was allowed to float in the latter part of the twentieth century, it was even illegal for an individual to own gold in Canada.

Nowadays, with gold hovering at $1600 oz, you can own gold, but you're still supposed to buy it, not help yourself. But let's face it – at $50/gram, a little bit of hi-grading is even more appealing.

Various methods are used to control hi-grading. Pretty much every province has producing gold mines and the police and private security firms have hi-grade cops whose job is to hang out in bars and social gatherings (undercover of course) and let it be known that they may be interested in buying gold; and also, of course, a yappy hi-grader always stands a chance of being ratted out by an envious or indignant co-worker. However, the main deterrent used by every gold mine is the "dry," and this is a dry story. It may be only a legend, but I am a sucker for legends. I will not name the mine or gold camp, but it could be anywhere. All precious metal mines use the dry system, and at every shift change the underground and mill workers pass through the dry on their way to and from their work area. We will deal with their passage on the way home.

The dry is actually split into two rooms, the work side and the street side, and in between is a shower room and an inspection area. The worker enters the first room, strips off his work duds, places everything in a metal basket which also has hooks for trousers, shirts, boots and whatever, and with a rope attached to an overhead pulley he hauls the basket up to the ceiling. The work duds are now out of the worker's reach and will be left to dry until his next shift, hence, "the dry."

The worker has deposited his lunch box at the inspection window and while he is showering the inspection staff will go through his lunch box, even dismantling his thermos, to check for gold. The employee then enters the "town" room, retrieves his town clothes from his locker and picks up his inspected lunch kit.

It's not a perfect system – no one is required to submit to a body-cavity search (cringe) but it certainly is a deterrent.

The Dry Attendant
(Credit for this part of the story goes to Claude "Ziggy" Jodoin, and everybody knows that Ziggy tells no lies??)

But in this great country of ours there is always room for entrepreneurs. And I've always figured this one fellow in particular was an entrepreneur bar none.

He was a dry attendant/maintenance man, a quiet, unassuming decent sort, always willing to do a favor or help a guy out. All drys have washing machines available in the work side to help the guys keep their work clothes clean, and while some guys were willing to hang around and wash their own clothes, most men would rather head home or maybe stop to get a cool one. The attendant would gladly wash your clothes for you for a small fee, and a lot of the workers were happy to make use of his services.

Now, the plot thickens. The underground workers are drilling, blasting and mucking rock, and their clothes get covered with dust, grime and mud, all gold-bearing. The surface workers are working in the crusher, ball mill and ore separation areas, and also collect minute particles of gold dust on their clothing.   Of course, the entrepreneur/attendant/maintenance guy knows this, and he modifies the washing machine drains, adding a series of traps in an area that he can access on his way home from work. Because he never works in the gold production circuit, he never has to do the dry routine. Periodically he cleans the traps and takes the sludge home. He is now an entrepreneur/attendant/ maintenance man/ gold-panning hi-grader, and for thirty years he lives an unostentatious life, his only visible luxury being his ability to give his kids an excellent post-graduate education. He retires and is never seen again – no doubt living the good life on the Cote d'Azure.

There are no flies on the company, though, no siree: They discover the trap system and institute a new company policy. They will henceforth wash your clothes at no charge, and they will clean their own traps, thank you very much.

I love that legend.

Scotty and the Hi-grade Cop
I know this story is true because I heard it from Sam, and Sam never told me any lies. Scotty is a wiry old gent and has been a fixture in the Red Lake area for as long as people can remember. He is a prospector, and like many of his contemporaries, he has had his share of finds, made a few bucks and has had his ups and downs. Easy come, easy go is Scotty's motto.

A hi-grade squad cop from Kenora is in town, and he is not exactly flying under the radar, if you get my drift. Whether it's because he is an annoying loudmouth, or maybe because Scotty is a canny old bugger, the cop's cover is soon blown. Scotty doesn't care – the cop has money to burn and he is buying the rounds - Scotty has some high grade for him.

Finally the cop figures he has Scotty roped in and he tells Scotty to meet in the parking lot for the transaction, and they do so, but they do not leave unobserved. Others have picked up on what is going on. They know Scotty, and they know Scotty has something up his sleeve. Very quietly, a few guys follow them out and crouch behind parked vehicles.

The cop tells Scotty he wants to see the hi-grade, so Scotty pulls out a well-worn, many-times-refolded claim map showing his block of claims on the East Bay of Red Lake – claims he's been trying to flog for years.

"There's your hi-grade," says Scotty, "Take your pick and shovel out there and you can high grade to your heart's content."

The cop is a big, burly guy, and Scotty weighs all of 120 lbs in a rainstorm. The cop picks him up and throws him clean over the roof of a '57 Ford. All the spectators emerge laughing, and the cop leaves town in a cloud of dust – no doubt to inspect parking meters until retirement. The boys pick Scotty up, brush him off and return to the Bucket where they will stand the drinks until closing time.

Chapter V   Fall of 1969

Mattabi Rush
In early September I have my "simple" operation and under doctor's orders take a full month off. Everything works out well – now I can jump into cold water any time I want.

October arrives and I grab a bus to Port Arthur. When I check into the office, things are in a bit of a tizzy. Mattagami Lake Mines has hit an ore body 40 miles north of Ignace, and a staking rush is shaping up. Mattagami Lake is a subsidiary of Noranda, and so far the news is pretty much in-house. As in Red Lake, Noranda has had McPhar fly the area, and now the airborne responses must be protected. Our contract stakers are busy elsewhere, so I am to take Rod and Lorne to Ignace and rail down some ground. We will only stake road-accessible claim blocks. The contract stakers will be on the job within a few days and they will hit the fly-in targets.

My faithful old GMC has been retired and a replacement is on order, so I rent a pickup. Me and the boys throw our packsacks and axes in the back and head for Ignace. On the way out of Port Arthur we swing by a snowmobile dealership and pick up an Amphi-Cat, the latest addition to our fleet. It is one weird looking contraption, sort of a plastic bathtub on wheels. It has a fibreglass body with barely enough room for three men on a couple of seats, and is powered by a skidoo-type motor. A constant velocity belt drive connects to a maze of bicycle chains transferring power to three axles, which protrude through supposedly watertight seals on either side of the tub. The Cat runs on six flotation tires, three on each side, and steers by two hand brakes a la caterpillar tractor or muskeg buggy.

Supposedly is the operative word here. Supposedly it floats on water, supposedly it is rugged and reliable, supposedly it will allow us to travel on old timber access roads, and supposedly it will save us time and long walks. None of the above is true. The Amphi-Cat will prove to be nothing but a big Amphi-Pain in the butt.

We hit Ignace and book into the Hi-way Motel. There isn't much action yet as it appears that we are first on the scene. We check out our first target. It is easy to get to with the pickup and we will start staking tomorrow morning.

After we eat I sit down with Rod and Lorne in their hotel room and go over the upcoming work. Neither of these boys have ever staked a claim, but they both have a summer's worth of work behind them and are familiar with the concept. They will spend the first day with me, and the following day will be running their own line. I decide to let them work together – I can't trust Lorne to work alone. He will help Rod, and I will work on my own. We will plan out the next day each evening, sorting out tags and arranging meeting spots. It's not rocket surgery, just plain hard work. Rod is on top of things and I'm looking forward to it. Staking can be fun.

I gravitate to the bar to see what is up and by golly, I spot Jim Lee. Jim is the Inco airborne supervisor I had first met at Moak Lake nine years ago. We are sure glad to meet again, so we share a round or two and catch up on the old days. I meet part of his crew and the only other familiar face is George Charity, the Anson pilot. They are working from a grass airstrip nearby and are as yet unaware of Mattagami Lake's discovery. They just happen to be flying the area, and I don't tell Jim much about what we are up to. It's not a big deal. Jim has been there, done that, and none of us are after secrets. We just enjoy rekindling old friendships and catching up on old news.

There is another guy hanging out in the bar and he is obviously enjoying himself. Jim tells me his name is Freddy Corcoran, and Jim thinks he is connected to Mattagami Lake. I will get to meet Freddy in a few days and will find we also have a bit of a connection. The exploration fraternity may be spread out across the Shield in Canada, but it's a small world after all – the Olympic Rings overlap.

The first claim block is a snap, and in four days we have the first 20 claims staked. On the last day we are walking back to the truck when a Whiskey Jack tries to land on Rod's shoulder, and Rod chases him off. He figures the bird is goofy, so I have to fill him in – so much to teach these young guys – so little time.

Sidebar: The Canada Jay /Grey Jay/Whiskey Jack/Camp Robber is a common sight in the boreal forest. They have little fear of humans, and like to hang out at tea fires looking for sandwich crusts. If you give a Whiskey Jack seven pieces of bread he/she will fly off in seven different directions to stash their goodies. We always doubt their ability to remember the location of every food cache. The squirrels probably eat the bread – squirrels don't miss much.

Rod had never seen a Whiskey Jack before. He had spent the summer at Black Beaver and had seldom packed a lunch into the bush. Also – Black Beaver was isolated from the main activity following Selco's find. Here at Ignace the birds know that hairless apes carry sandwiches.

Our next claim block is quite a ways in on an old tote road. The area had been cut over 20-plus years before, and many old roads are semi-passable. Air photos tell us where these roads are, but air photos don't tell us the whole story. Therefore, we carry the Amphi-Cat in the pickup box.

So today we find the road blocked by a beaver dam five miles from our staking target. The dam is built alongside the road, but it's an old dam, has breached in one spot allowing water through, and the running water has cut a channel across the road. We study the situation. Vehicle tracks tell us that someone has made it across the gully recently, but we figure it was a four-wheel drive unit. It looks a little iffy for our two-wheel drive Ford so we unload the Cat. We don't expect to run into competing stakers – the real rush is not under way yet. Moose season is open, so our company in the bush is more likely to be hunters.

With the Amphi-Cat on the ground and raring to go we decide to take advantage of the beaver pond to test this little bathtub's ocean-going capability, so Rod and I climb in and drive into the pond. Well, what a farce! We have barely two inches of freeboard and we don't dare to shift our weight much. Three men will surely sink it. The cat has no propulsion system, the cleats on the floatation tires are supposedly going to move us through the water, but it barely inches along. The dam is full of old branches from the beaver days, and all it takes is a little twig to hang us up, tires spinning away happily with us two goofballs in the tub, hoping no one else comes along to laugh at us, we are doing enough laughing at ourselves. We even have to help this piece of crap out of the pond. Amphi-Cat will stick to dry land from now on.

The three of us head into the claim block perched on the six-wheeled contraption, and four miles in we pass two parked vehicles. Hunters in an Oldsmobile and a Cadillac have crossed that washout! Now we feel really foolish. We continue on to our day's work and on our way out that afternoon one of the bicycle chains breaks in the bowels of the Beast. We have to walk the last 3 miles.

The next morning we corduroy the washout with lengths of poplar and drive in to the claim block. On the way out we toss the Amphi-Cat into the back of our pickup and there it will remain, an expensive hunk of ballast.

A few weeks later I return the pickup and Amphi-Cat to Port Arthur and drive back in my new GMC panel. At the P.A. office I ask Don "Who had the bright idea to buy that hunk of excrement?"

Unbeknownst to me, Ralph is lurking in a adjacent office and he pops up and says, "That would be me," Oops!

(The rolling bathtub disappeared and Peter told me Ralph had taken it home for his boys to play with. I always wondered if Ralph had shown that item on the company books with three little words added, "Bury as meat.")

Table Scrap

Not too long ago, Rod and I spent an hour or two on my back deck. It had been 40 years since we had talked and we had a lot of catching up to do.

Ralph entered the conversation. Rod asked me if I recalled "the stump" at Black beaver. I did not, and he filled me in.

They had felled a large jackpine between the cook tent and the dock, leaving a four-foot high stump. They had used their axes to make a smooth face on one side and had drawn a head and torso likeness of Ralph thereon. Then, on an evening, as a cabin fever reliever, they would take turns throwing axes at Ralph. This shocked me a little – I had thought that I was the only one with negative feelings for the guy. Poor Ralph – he tended to polarize people he dealt with.

I know why I forgot that. I didn't ask Rod, nor do I want to know – Maybe my picture was on the other side of the stump.

The news is out now about Mattagami Lake's find, and the bush is heating up. Our contract stakers arrive with John and KHM, and they are hitting the fly-in targets. Most of the action is east of Hwy 599, and on up to Sturgeon Lake. Rod, Lorne and I still don't have much competition on our side, and it's a good thing we don't. Our walks are long, and we are pretty darn slow compared to the contractors.

Bill and Dave are our contract stakers, and John moves their camp every two days, but before they head into the bush, Bill tells me something that shocks me to my boot tops. It seems that one of Jim Lee's men has let it be known that copies of Inco's airborne maps are available for a cash donation to his retirement fund. Now, I'm not a goody-two-shoes, nor do I feel I owe Inco anything, but darn it all, some things you just don't do! I figure you ride for the brand, and if you don't like the outfit then quit and find another job. I worry about this for a couple of days, turning the problem this way and that in my head while I study the angles. Finally, with my mind made up, I go to a pay phone and make a person-to-person collect call to Sudbury. I want to speak privately to John Mullock, the project manager at Contwoyto in 1963. He is the only one I can trust to handle this with a cool head. I don't tell him who the double agent is, John can figure that out for himself, but I make sure John understands three things: The first is that Jim Lee is in no way involved. Secondly, no one in the Noranda camp would stoop so low as to purchase any stolen info, and thirdly, I do not expect any compensation for ratting someone out. In fact I ask for and receive a promise of anonymity.

Two days later, ten Inco men arrive at Ignace and proceed to protect their anomaly responses with blocks of claims. (I never did tell Don about my phone call. I figured that was between me, John and my conscience. John did reward me two years later with a couple of staking contracts. Fair enough, I figured.)

Sidebar: It turned out that Freddy Corcoran wass a partner with Jack Hodge and Garth Zimmer, the two guys I had met back in 1965 on Nighthawk Lake during the Texas Gulf rush. When Mattagami Lake hit their glory hole, the project manager realized that he needed to solidify his own ground position, so he called Jack and Garth in to stake additional ground in the area of the main strike. Freddy is their town liaison, coordinating camp moves, grub supplies, and recording claims. I don't get to see Jack and Garth this time – they have their noses to the grindstone in the bush. I will run into Jack in Red Lake next January when the Olympic Rings will interlock again.

We rookies are done staking. Noranda has all the ground it wants west of 599 and we can go home. It's almost December now and has been snowing a lot. Our last week has been heavy slugging on snowshoes in ever-deepening fresh snow.
Before we leave, Don and I have a chat. If we want to keep staking on our own, Don will pick out a few areas and back us with rooms and meals for a 10% interest. I point out that open ground north and west of Ignace is just so much moose pasture, but Don feels anything will be of interest to junior companies. It sounds like a sweetheart deal and I talk it over with Rod and Lorne, and they are not interested. They want to go home – I can hardly blame them - they have been in the bush since June. Perhaps I should have sold it a bit more aggressively, but I admit I am also eager to see my family, so I pass the word on to Don. "Thanks but no thanks." Rod, Lorne and I leave the next day.

(When I returned to the area in February, every sliver of land was staked solid from 599 to Sioux Lookout. Woulda, shoulda, coulda – the door to opportunity had opened and we walked right on past. Such is life.)

It's nice to be home in December. My girls are still pre-school so we get to watch Mr. Dressup and Chéz Helène and I get to tuck them in at night. The tree has to be cut and trimmed, I spend time with my parents, visit old friends and make some preparations for the upcoming season.

In early December Chic and Leo come to Fort Frances. A few years before, they had discovered a copper showing east of Rainy Lake, and a follow-up program with Sam's drill had proved up a relatively small high grade copper deposit. It was too small to interest Noranda so they had spun the property off to a couple of Port Arthur promoters who in turn had formed a company to develop the mine. Don told me that it would take only five years or so to extract the ore, but the grade was sufficient to show a profit for a junior company. Chic and Leo were in town for a bit of a signing ceremony. They would receive some cash, of course, and with some stock options and participation by way of a 3% net smelter return (NSR,) there could be some ongoing financial benefits. I thought it was pretty cool that these two old guys would see some tangible results of their prospecting endeavors.
So Chic and Leo are in town along with Don, Ralph, Sam and the two Port Arthur promoters. I am not part of the meeting (signing ceremony) but I go to the bar that night to join everyone for a few celebratory drinks. I am sitting with Sam and Ralph when Don comes to the table and he is obviously upset. He has been upstairs with Chic and Leo and Don tells us that the two old farts have been in a fight. I guess there had been a disagreement over how the deal had been handled and the argument had escalated to a punch out. Don had separated them and had put Chic into his (Don's) room. Don asks me to drive Chic to PA the next day. He tells me to take my wife with me and do some Christmas shopping, so the next day we take Chic to the Fort William airport.
Chic doesn't say much during the drive down. He is badly hung over but once in a while he mutters a bit. "I drew blood" says Chic. "I hope I broke the old bastard's nose."
What a way to end a partnership spanning more than 20 years – pretty sad.

Table Scrap – Chic and Leo

Chic and Leo retired from the game that year. Never again would they haul the Taj Mahal into the Shield, but I sure hope they patched things up on a personal level. The thing was – they had been partners for <u>more</u> than 20 years. They had quite a back-story and as usual, I pieced it together by way of common room chat-fests.

They were both from the Ottawa Valley and had hammered rocks independently before the war. They enlisted, ended up in the same outfit and shook hands on a partnership deal while ducking bullets in France. When they mustered out they hit the bush and three years later joined Noranda.

They both lived in Ottawa in the off season – separately of course – only one was married. Winters were spent studying maps and planning the next, sure-to-be-exciting summer. Maybe this year they would hit paydirt!

As they got older they slowed down, and Leo began to lose his eyesight – close up was fine – distance became a problem. So for the last three or more years Chic had been on double-vision duty.

On the claim line Chic ran the compass and Leo followed with the axe. On the showings Leo never strayed far. All handles on the shovels, axes, picks and grub hoes were painted day-glo orange – thus when Leo laid down a tool he could find it again. When trenching and blasting, just as in France, Chic had Leo's back.

Old partnerships, like old marriages, should mellow out – sit around, remember the good times and get a chuckle out of that minor punch-up in Fort Frances.

So that's why I hope Chic and Leo patched things up.

I have a peripheral connection to Chic and Leo's copper find. Back in July, Don had told me that a junior company was being formed to develop the mine, and Don was going to pick up a substantial piece of the initial public stock offering (IPO). The initial price is to be 30 cents a share, and if I want in Don will swing 10,000 shares my way. I like the idea of playing with the big boys, so I say, "Sure." I trust Don and in a way I guess Don is offering me a bit of a reward for a job well done. I go to my bank manager, borrow 3000 bucks and prepare to enter the world of high rollers.

You have to put this into context of 1969 dollars: $3000 was no small potatoes to a family man earning $500 a month. I could buy a decent pickup for that kind of money, and in today's dollars you can add a zero. But the loan was at 6% and as it was a demand loan I only had to pay the monthly interest. The principal can wait until old money bags me cashes in at 2 bucks – yeah, right.

(Says the English Professor: "Although while certain instances may justify a double negative, there is no such thing as a double positive."
Says a student, "Yeah, right.")

So I watch and wait with my 3000 bucks sitting in my bank account. No stock appears and I start to bug Don about it. "It's coming, it's coming, hold your horses!" So I hold on.

In August, Seemar hits the market at 33 cents. So far, so good, but I still have no stock. Within two weeks Seemar is trading at 22, and I receive my 10,000 shares. I could have paid

$2200 instead of $3000. I am a bit pissed, and I tell Don so. "It'll come back," says Don, and I think he is a little sorry by now that he has tried to do me a favor.

But Don is right. The stock starts to climb and by the time I am sitting in the bar with Ralph and Sam, and while Chic and Leo are punching it out upstairs, the stock is at 71 cents. I want to cash in – $7100 is a pretty good haul of moola. I ask Ralph's opinion – I know he has at least 100,000 shares and I want to know what I should do. Ralph says hang on, the stock will soon break the dollar mark and the sky's the limit, so I hang on.

Now here's the kicker. While that mealy-mouthed slime ball is telling me to hang on he has a sell in at 75 cents. The stock rises to 77 cents two days later and rolls right off the end of the table. I try to sell at 70 but the stock goes down to 60. As fast as I lower my sell the stock drops even faster. At 30 cents I give up and by January it hits 11 cents. I've learned a hard lesson. Minnows shouldn't swim with the big fishes, especially with Ralph the Shark.

However, my foray into high finance is semi-successful; almost 2 years later I return from a small contract in Red Lake and pick up a copy of a financial publication. I check out the penny stocks, and find that my horse has regained a little life. The stock is at 33 cents! I phone my bank, put in a sell order, and when the smoke clears I have paid the loan and brokerage fees with about 10 bucks left as profit. Whew! Forget the $360 interest I have paid. I consider it tuition.

Before Christmas Don calls to tell me I am going back to Black Beaver in January to fill in empty spots on the grid, now that the beaver pond is frozen. I will have KHM at my disposal and we need a pilot. Can I find one? I ask my old high school bush pilot buddy and he tells me Fred Kozik is available. I look up Fred, describe the job and the compensation package and he is interested. I call Don and he asks me how many hours the guy has. Jeez - I never thought to ask. I figure my friend's reference is good enough.

I phone Fred, "How many hours in your log book?"

"Pretty close to 8000," he says.

He is hired on the spot.

Chapter VI   1970: A Mix & Match Year

Back to Black Beaver
January 2nd and I am on the road again. Rod, Lorne and I are on our way to Red Lake where we will mobilize for a return to the camp at Black Beaver. We arrive in Red Lake at 3 pm, check into the Red Lake Inn, and while Rod and Lorne head to the Cash and Carry to put our grub order together I go down to OCA to say hello to Dave and arrange some flying. We need to take in stove oil and avgas in addition to our normal supplies. Fred and KHM will be joining us shortly, but OCA will be doing the heavy lifting. We will be leaving Black Beaver before the end of the month, but Sam's drillers will be using the camp after we leave. They have two or three drill targets on the grid.

We have our supper at the Inn and when I walk into the dining room, there is Jack Hodge sitting there, so of course I join him. I have not seen Jack since Nighthawk Lake four years ago and we catch up on the intervening years. Jack has a man with him and is doing some geophysics on a couple of relatively small claim blocks near Red Lake. He can drive to his jobs and is working out of the Inn. His partner Garth is working in Northern Quebec and Jack will join him in February.

I tell Jack I was in Ignace last fall, and he fills me in on the Freddy Corcoran connection. It seems Freddy had an inside track with Mattagami Lake and he had brought Jack and Garth in on the staking contract. When Mattagami Lake had picked up all the ground they wanted a narrow strip remained open, and although the project manager felt it was a good bet, the head office refused to stake it. Before Jack and Garth left Ignace they put in a block of claims for themselves, and had already dealt the ground to New Brunswick Uranium (NBU). Jack didn't know it yet, but by the time the smoke cleared a couple of years later, there would be three new millionaires on the Shield.

The next day it's snowing and the aircraft are grounded. No problem, we still have work to do. Among other things I rent two oil stoves and take them down to OCA.

We leave for Black Beaver bright and early the next morning. There is now almost three feet of snow in the bush and I hope the tents are in good shape. As we circle the camp before landing everything seems fine and dandy. We land on two feet of lake snow and there is no slush evident.

The camp looks fine from the outside. There is quite a bit of snow on the tent flys, but they had been well braced last September. We will fabricate a roof rake and pull the snow down.

We pull the nails holding the tent doors closed and find the bunk tent undisturbed with the mattresses rolled up on the steel cots, just as they had been left. The kitchen, however, is a disaster! Squirrels had chewed a small hole near the peak at the back wall and they had vandalized the joint. Every roll of paper towel and toilet paper has been shredded into pieces like so much confetti. Half-empty cereal boxes had been left behind, and after eating the contents, they had also shredded the boxes. The Froot Loops and Count Chocula must have given those destructive little buggers quite a sugar high, and it looks like it was one hell of a party. As if the mess isn't bad enough, every flat surface is covered with frozen squirrel pee and poop. Well what can you do? I walk outside and fire every squirrel within a two-mile radius, but I doubt they hear me. They are most likely in semi-hibernation, farting Rice Krispies.

So I light the oven on the propane range for a little interim heat, and while Rod and Lorne take out the air-tights and clear the snow off the tent roofs, I chop a water hole in the ice and put a couple of pails of water on the stove. Then I begin to clean up squirrel excrement. First I give the kitchen table a good scrubbing, then I wash every dish, cup, bowl, pot, frying pan and piece of cutlery, and pile them on the kitchen table. Then comes the shelves and counter top – you get the picture. The oil stoves are brought in and I take some time to direct Rod and Lorne regarding their installation and go back to the dishpan.

It takes me all day, but by sundown at 4:30 the floor is washed, everything is back on the shelves and we have our supper in a nice, clean, warm kitchen. Tomorrow we will give the camp a final spiff up and prepare to do our geophysical work. I set the kitchen oil stove at 1/4 throttle and we hit the sack, tired but satisfied with our day's work.

Bright and early the next morning I hop out of bed and boogie to the kitchen to put the coffee on. I open the door and, Holy Old Double Decked Dinah! What a sight! The oil stove has backed up overnight and the whole shebang is covered in a layer of black, oily soot. Every horizontal surface, every dish, and as an added attraction, all the new groceries are covered with this crap. I have to clean everything again.

We can't have breakfast yet. I wash the coffee pot and three cups. I check the oil stove firebox and it's plugged with soot. When Rod and Lorne show up I give them a cup of coffee and tell them to dig the old airtight out of storage. The oil stove is heading for the snow bank.

(I have a problem with oil stoves, mainly because I don't have a clue regarding their care and maintenance. In my world, oil stoves are a 50/50 deal. Either they work or they don't work. Thus the bunk tent is toasty, warm and clean, and the kitchen is not!)

I start washing while Rod and Lorne bang the soot out of the stovepipes and set up the airtight. It's a good thing I had the foresight to have a pilot pick me up additional cleaning supplies yesterday. I need them, as I wash everything three times, while the boys cut a supply of firewood. By noon we have our bacon and eggs, by nightfall our kitchen is clean again, and tomorrow we can actually do what we came here for.

It takes less than a week to finish the Black Beaver grid and I catch a back haul to Red Lake. Fred and KHM are coming in and they will stay in camp with us for two weeks. We have two or three anomalies to check out and will fly daily from camp. I call Don and find out that Fred will be a day late, so I get to spend a couple of evenings with Jack Hodge and Ray Willet, his helper. Ray is from Val D'Or and Jack tells me he is a good man, but when Jack finishes his Red Lake contract he will have no work for Ray. H'mm…. I file this info for future reference. We can always use a good man.

Two days later Fred arrives. We load up KHM with groceries, take off for Black Beaver and it seems the blue and white 180 actually wants to fly. The takeoff run is pretty much normal and as we gain altitude I ask Fred what gives. Has KHM had some airframe adjustment? He gives me a raised eyebrow so I tell him how poorly the Cessna had performed last summer. I also tell him John's theory. John had said that KHM had hit the bush two years ago and he figured that during the rebuild that the wings had been set up with a slightly askew angle of attack, requiring a higher speed to lift off. Fred has two words for John's theory and the initials are B and S. "You fly the plane" says Fred, "never let the plane fly you."

It doesn't take long for Fred to set me back on the right path. The next morning, after breakfast he says he is going to put some avgas into KHM before we leave, and I tell him Rod will help him. "I fuel my own plane," says Fred and he storms out.

Now remember, I had spent last summer baby-sitting this aircraft, and I am a little slow on the uptake. By the time Fred has fueled KHM we are ready for the bush, so I pull out my maps prepared to discuss things with him. Our target is on a fairly small lake, and does he want to put us in there? And shall we walk out to a larger lake nearby for our afternoon pickup? And yada-yada?

Fred glowers at me and says, "Look here! I fly the plane and I decide whether or not the lake is big enough! Let go of my ears!"

Whew, what a cranky guy! I don't mind though – I've got a bush pilot now.

(One more thing about Fred. Don't ever let him catch you taking a dill pickle out of the jar with your fingers. You'll get your knuckles rapped for sure.)

Towards the end of January our work here is finished. We turn the camp over to Sam's drillers, already on their way. KHM left yesterday for Thunder Bay so we catch one of Sam's empty OCA planes on a back haul to Red Lake. The three of us will head home for a couple of days off. Lorne is quitting. He doesn't like life on the snowshoe trail.

In Red Lake I again run into Jack and Ray who are also almost finished their contracts. I hire Ray, and Jack will drop him off at Kenora in a few days. I spend two days at home, make a swift trip to Thunder Bay for camp gear, maps, and instructions, and return for one more night at home. The next day Rod and I will head for Kenora where we will take part in a 10-day IP survey at Squaw Lake.

Sidebar: A heads up to avoid place-name confusion: In late '69 Port Arthur and Fort William were amalgamated. A referendum was held to choose a name for the new city. The voters were given three choices: "Lakehead," "The Lakehead," and "Thunder Bay." Popular consensus at the time was that the traditional nickname, "The Lakehead," would be appropriate, but the consensus on City Council was in favour of "Thunder Bay." By offering two "Lakehead" choices, the vote was split, and "Thunder Bay" emerged the clear winner. On January 1st 1970, the name was officially changed to "Thunder Bay," and will remain thus for the rest of this story. (City council was right. Thunder Bay is a dandy and more appropriate name.)

Table Scrap

Sam's drillers flew a John Deere cat in to Black Beaver (as described in winter of 62 Chibougamau) and while moving the drill across the beaver pond west of camp the cat went through the ice. Beaver pond ice can be tricky, as the beavers will lower the pond water level in winter, probably to allow some air space under the shore ice. The ice will be ok in the centre of the pond, but near shore you have to be careful. The water was not deep where the cat went through, but Vincent, the operator, was thoroughly soaked so he boogied back to camp for dry clothes.

When you sink a piece of machinery in the wintertime you don't mess around. It's imperative that you get it out of the water before it freezes in, so you quickly rustle up three large trees for a tripod, attach a chain or length of cable to the cat and hoist it out of the water with a set of chain blocks. When the cat is clear of the water you drain the oil and fuel and leave it overnight. The open water freezes and the next day you slide a few poles underneath for added strength, refill the drained crank case and fuel tank and drive it away.

That is how it is supposed to work, but when Vincent returned in his dry clothes he jumped on to the suspended cat, the cable broke and John Deere went for another swim taking Vincent with him. The boys fished him out again, told him to stay in camp the rest of the day and laughed like crazy.

You might say Vincent was accident prone. A week later he was splitting some kindling wood in camp and he cut his finger off between the first and second knuckle. An OCA Beaver happened to be unloading diesel fuel at the time, so Vincent was hustled down to the plane. Someone said that perhaps they should send the severed digit back with Vincent, but as the guy ran back to the chopping block a raven swooped down and flew off with Vincent's finger. I figure it may have been the same bird that bent Nick's baseline at Bridges Township – those ravens are a mischievous bunch of birds. Anyway, Vincent can now order half a cup of coffee a la Raymond Paradis.

Squaw Lake

Rod and I pull into Kenora and check in at the Northland Hotel where Ray is waiting. The Scintrex man will arrive tomorrow; meanwhile, we have some mobilizing to do. We have to visit Parsons' airbase and Fife's Hardware.

Fife's hardware has been in Kenora forever. I don't know when Fife's started, but it must have been back in the day when Kenora was still Rat Portage The building is a three-storey wood and brick structure, and what a treasure trove! Fife's is the epitome of the phrase "If we don't have it, you don't need it," but in Fife's case you can add, "We will get it for you." (Kenora is a compilation of the names of three merged old-time towns – "Ke" for Keewatin, "no" for Norman and "ra" for Rat Portage.)

On the main floor an old wooden display counter on your right runs the full length of the store, and behind the glass front are hundreds of items – hunting knives, jackknives, compasses, jeweler's loupes (loupes are part of every prospectors ditty bag) – far too many items to list. Behind the counter, shelves reach to the top of the 12-foot ceiling. A little brass rail is recessed in the floor, and a ladder with little brass wheels can be pushed back and forth the length of the store, supported near the top of the shelves by a corresponding brass rail and another set of wheels. The rest of the main floor is full of display islands, and on the walls hang shovels, scythes, pickaxes – too much, too much, too much! Do you need an axe? You have hundreds to choose from, anything from little camp hatchets to six-pound wood splitters. They even have huge double bitted New Brunswick axes, as well as broadaxes for squaring timbers.

This is great. I pull out my shopping list and I need some weird stuff. I think maybe I can stump Mr. Fife, but he is unstumpable.

Some of the supplies I need are pretty basic – shovels, a grub hoe and a small toboggan. Now comes the hard part. I need more than 5000 feet of copper snare wire (aka rabbit wire) and any other store that carries it sells it in little 20' coils. No problem. Fife's has 1000 foot spools of rabbit wire. Ok, now I need salt, lots and lots of salt. Water softener salt will not do the job for me, and grocery store salt is too expensive. I need the type that the Highway Dept. uses on icy roads – and they buy their salt by the truck load. No problem, Mr. Fife can get me road salt in 10-pound bags. I order 50 bags to be delivered to the airbase in two days. I also need aluminium foil, and household-sized foil wrap just won't suit us. Again, it's no problem. Fife's has food service-type foil in big rolls – is there anything this guy doesn't sell?!

But now comes the kicker, and I think I've got him this time! Does Mr. Fife have bulk felt? (We will be sleeping on spruce boughs, and I want something with some insulation value to put under our eiderdowns.) Mr. Fife says "Come with me," and we head upstairs to the third floor. As we pass the second floor I see snowshoes, canoes and tents, and the third floor is a step back in time, a virtual museum! Hanging on one wall is a double oxen yoke, and on another wall hangs two complete sets of work horse harness, from collars and hames, right back to breeches and tug lines, all shiny black leather with silver stud adornments. I see single trees, double trees, fancy bridles, rolls of reins, and even a couple of buggy whips, and to top it all off, a single horse buggy with red wheels sits in a corner! I surreptitiously wipe some drool from my chin. This is flipping amazing!

Mr. Fife leads me to a huge roll of 5/8" thick felt 8' wide. He will cut off 4 pieces 3'x 8' and I say, "Deal!" I know Don will bark at the price, but I don't care! I want everything! I want it all!

I get a bonus when I buy the felt. I have size 15 feet, and I can never find felt insoles big enough for my moccasins. I trim one foot off each felt mattress and I have enough material to keep my feet warm for years to come.

(In 2002 I am working at a tourist camp on the northwest part of Lake of the Woods, 40 water miles from Kenora, when I hear that Fife's is closing the store. Winnipeg is only two hours away and the big box stores are killing the family operations. Fife's is having an auction, and I can't attend, but I'm sure there were many deep pockets from all across North America at that auction.)

I'll use up some space here to explain the mystery of some of my purchases. We are going to do an IP (induced potential) survey on the Squaw Lake grid, and the theory behind IP as explained to me follows.

Not all commercially viable ore bodies are of the massive sulphide type. If the minerals are disseminated in the host rock, the lack of continuity will render an electro-magnetic survey pretty much useless. No continuity – no conductor, therefore an IP survey is necessary.

The IP operator sets up at a grid station and induces an electrical charge into the overburden or outcrop. Receiving stations are set up further along the picket line at 100 foot intervals. Usually three or four stations are used and light insulated wires lead back to the transmitter/receiver. The operator fires up his little motor/generator unit, and takes his first readings, to measure resistivity, indicating how much of the induced current is reaching his remote locations. (The insulated wires are connected to the receiver unit and each station can be read individually.) He then shuts the motor off and takes his dwell readings at each station. Think big battery – the dwell indicates the potential of the disseminated sulphides to hold a charge. The shorter the dwell time, the less metal in the host rock, ergo, induced potential.

I'll leave it at that before you become as confused as I am. I can already see your eyes glazing over and I can hear you humming a little tune. Trust me, brains larger than ours can process this info back in Toronto.

(Incidentally, a disseminated ore body is by nature a low-grade type, and is usually an open pit mine. Underground mining is far too expensive to be economically feasible.)

Now back to the reasons for the Fife's Hardware buying spree. For the bush stations we have to have good contact with the overburden, and the ground is frozen, so we shovel the snow to clear a decent sized area to work in. We carry a two-burner Coleman stove with us and melt snow in a large pot while adding salt to create a strong brine. Then we pour the hot brine on the ground thawing it down to 8 inches or so below the surface, digging out the now melted dirt/gravel with our grub hoe. We then plant a copper alloy stake (electrode) and pack it with brine soaked dirt/gravel. We now have a good contact with the earth and the brine will also prevent freezing should we want to reuse the station later.

If the station is on outcrop we make a little dish out of aluminium foil. We set a little copper pot/electrode in the foil packet and fill the basket with hot brine. This is all very labour intensive stuff, but necessary for a good survey.

Rod and Ray will handle the bush work and help the I.P. guy string the wires. My job is on the lake.

The grid had already been cut by contractors after freeze-up. The survey is to be six line miles, and half of these miles are on the ice of Squaw Lake. I am going to auger more than 150 holes through 30 inches of ice and set up the I.P. stations. At each hole I will wrap a softball sized rock with tin foil, wrap copper snare wire around the tin foil, and lower the rock to the lake floor. Then I wrap the wire around the picket on the lake surface, and cut the wire off, leaving enough extra to hook up to the transmitter/receiver. It's not really all that complicated, it's just hard work. This stuff doesn't take much brain power – just muscle.

With everything organized we take our camp gear down to Parsons and fly in to Squaw Lake. Our camp will be a snowbank set up. We only plan to be here 10 days. However, plans don't always work out do they? It will be three weeks before we see the last of Squaw Lake.

Our plane is a brand spanking new Cessna 185. It sits there all shiny in the early morning sunlight, and we hesitate to even touch it much less throw in our grubby gear and climb in with work clothes on. As we take off, the North Country seems a lot brighter to me. In fact it is a lot brighter! The Cessna windshield and side windows are plexiglass and after years of scraping frost on cold winter mornings the windows are usually marked with little scratches and cloudy spots, sort of like a car windshield sand blasted by years of driving on country roads. These windows are crystal clear, and it's like your first trip in dad's new car – not a rattle, and it even has that new car smell!

Sidebar: The Cessna 180/185, unlike the DeHavilland Beaver/Otter were not specifically designed for northern bush flying, yet they fit into the bush plane mold. Their only drawback is that after years of rough water and frozen snowdrifts, everything gets a bit loosey-goosey. Instrument panels and interior trim start to shake and rattle, and over time the rivets on the metal skin will loosen somewhat. (Handy hint: If you're buying a used Cessna, check the logbook – if it shows a lot of hours on floats/skis north of the 49th parallel, you might want to pass on it.)

Squaw Lake is at the east end of the Northern Peninsula, a strip of land extending into Lake of the Woods from Manitoba, and is only 20 miles from Kenora. Two short hops later, and our camp gear is on the ice. We will build the camp today and Bud Parsons will pick me up this afternoon. Tomorrow I will meet the Scintrex guy at the Kenora airport.

It doesn't take us long to set up camp, a tent in a snowbank is just one step up from a refrigerator box on a sidewalk heat exhaust.

First we shovel snow. Our 12x14 light canvas tent will be pitched a la summer style, and we clear a 12x14 perimeter. We don't shovel all the snow out, though. We leave the snow at the rear and clear the front 6' or 7' to the ground. We set the tent up with squaw poles and put logs around the edges to hold down the skirting. Then we build a 2-foot log wall crossways in the middle of the tent and pack the snow behind the wall with our snowshoes, adding snow to bring it up more or less level with the log wall. We cut spruce boughs and lay them on the snow behind the wall until we have a nice springy three-inch layer to put our felt mattresses on. The front six or seven feet cleared down to the ground will be where we set up the airtight heater and a very, very basic counter to cook on. We will sleep side by side on the 7x12 foot platform like 4 pigs in a blanket, trying to adjust that pesky branch digging into our ribs without disturbing our next door neighbour, and hoping to avoid the call of nature before dawn. We will eat our meals sitting on the log wall with our plates on our knees. This is most definitely not the Beverly Hills Hilton!

The next day I pick up the Scintrex guy and the I.P. unit at the Kenora airport and we head back to Squaw Lake. The I.P. guy is unimpressed with his home away from home, and I can hardly blame him, but he is a pretty good guy, and he won't be here long will he? (Yes he will!)

The next morning we hit the job bright and early, and by noon the first electrodes are planted and the darned I.P. unit won't work! The guy tries everything he can think of, and although the unit is brand new, it just won't produce a charge. He will have to call Toronto.

I have my own little problem. I had brought an old hand auger from Thunder Bay, and it won't cut ice. I could make more headway with a sharp tablespoon. I will have to go to plan B.

We have no radio in camp, but Bud flies over almost every day and we have arranged a signal system. We make a spruce bough cross on the Lake and soon Bud spots it on a back haul and drops in. It is already late Friday afternoon, and since we don't expect to return with a new unit before Tuesday morning, I take Rod and Ray back to Kenora with us. They can't pre-set electrodes, and I don't expect them to cool their heels at the Squaw Lake Hilton while I am living La Vida Loca in Kenora.

In town I call Don and fill him in on developments. He's cool – after all, crap happens. In an uncharacteristic fit of magnanimity, Don tells me to take everyone to the new Holiday Inn. He will call to arrange the company credit card payment.

This is pretty fancy digs for us bushwhackers. The Holiday Inn is a new seven storey circular hotel built on the Kenora waterfront beside the mouth of Lauritson Creek where it empties into Commissioner's Bay, and has only been open for three months. A central elevator takes us up to our pie-shaped rooms. The dining room and cocktail lounge are on the top floor, giving the customers a panoramic view of the bright lights of Kenora and the hustle and bustle of Commissioner's Bay as Skidoos travel this way and that. In the summer time the bay would be full of boats of every conceivable size and type, dodging floatplanes. On the far side of Lauritson Creek is a Safeway grocery store with 15 or 20 boat slips. You can tie up your Lund 200 horse Magnum, do your shopping, wheel the cart down to the boat and return to your summer home. This is unique in Canada and possibly in North America. Pretty neat eh?

We grab a shower, put on our cleanest dirty clothes and hit the dining room. The menu shocks us. These are 7th storey prices for sure. Ray tries the steak tartare, eats half of it and saves the rest (just in case he wears a hole in his boots). I order the Orange Roughy. I don't have a clue what Orange Roughy is, but I like oranges, and the name intrigues me. Turns out to be a flat piece of tasteless fish – pretty thin soup compared to pickerel. Rod can't find a hamburger and fries on the menu, so he settles on a T-bone steak. "Not bad," says Rod, but he thinks they should have let the poor little calf grow some before they butchered it. He also wishes he had ordered baked potato instead of mashed, then he might have actually got a whole potato.

We are starting to get a bit giddy. We aren't so ignorant as to laugh out loud, but the giggles are coming, so we decide to cool our jets in the cocktail lounge. We are early arrivals, and we choose a table in the centre of the room and order a beer, and although the beer is domestic, the price is imported. Now more people start to come in; men in suits, women wearing fancy dresses and long gowns. They don't let on that they see us, but we notice they choose their spots around the perimeter of the room. We are protected by an invisible force field, and we are laughing out loud by now. We wonder who do these jerks thing they are, and what's the big deal about this joint anyway? It doesn't even have a shuffleboard! We go to bed.

Saturday morning we move to the Northland Hotel – home at last! The Scintrex guy stays at the Holiday Inn. He's not as snobbish as us yahoos.

94

Sidebar: In a few years the Holiday will open a regular bar on the main floor and will it become diametrically opposed to the 7th floor lounge, with knife fights at the Ok corral every weekend. I'll take a frontier bar any day. They may have their rough edges, but they tend to be more convivial. In fact, the Bayview Hotel in Vermilion Bay had a rule they enforced for years. Leave your knife with the bartender and enjoy your beer, boys!

I know Don's offer to put us up at the Holiday is an attempt to stroke my ruffled feathers. It is clear to me now that with Peter on board that I will become nothing more than a glorified party leader, and Peter is already running things east of Thunder Bay. On my last trip to the office I had told Don that I needed a raise and I had set the bar high enough to make sure I would sink or swim. I figure I will sink and have already had some discussion with Sam about going to work on his drills. We will see.

The new I.P. unit will arrive shortly after lunch, so in the morning I go to Fife's. I don't relish pulling a power auger around on the lake but I guess I'll have to. Maybe Mr. Fife will know where I can rent one.

"I have just what you need," he says, and hands me a contraption that looks like a large pogo stick except that instead of a spring at the bottom there are two cutting edges with half-flight of augur on each. "This is from Sweden" says Mr. Fife, "And here is the instruction manual and the tool kit." He grins and gives me a 4" by 6" piece of paper and a 6" nail. (I'm telling you, the more I see of this guy the better I like him.) On the piece of paper is one sentence. "In case of dullness stroke the sharpener twice along the upper edge of the cutting surface." Criminy – this is low tech for sure! I have my doubts but Mr. Fife hasn't steered me wrong yet. I put the tool kit in my pocket and take the Sandvick auger with me. I'll try anything once.

We pick up the new unit after lunch and by 4 pm we are in camp. We dig our potatoes and steak out of the snow bank refrigerator, have supper and hit the sack. We will be in the bush at 7 am tomorrow morning.

The next morning I haul my packsack load of rocks, tin foil and copper wire along with my shovel and ice auger out to the lake and once again I am amazed, amazed. That Sandvick pogo stick goes through the ice like a hot knife through butter. Every six inches or so I give the cutting edges 2 strokes with my spike and the ice chips fly. It takes me longer to shovel off a four foot square section of snow than it takes me to drill through 2 ½ feet of ice. Thank you Mr. Fife.

All is not well with the rest of crew by noon Scintrex Guy realizes he has another dud on his hands. This time we don't mess around. The spruce bough X goes out, Bud comes in and its supper at the Northland.

I phone Don and give him another tale of woe. There's nothing we can do about it says Don, and he's dead on – we do nothing for seven days.

It turns out that this unit is of a new, improved, design and there are some glitches at the Scintrex R and D department. Tomorrow they say, and tomorrow never comes. I'm getting cranky with the I.P. guy, but it's not his fault. I just wish I had gone home for a few days. Home is just two hours away and the Northland is losing its charm. It's also hard on the wallet.

Finally we get a reliable unit. Back to Squaw lake, and we are running between stations now. There is nothing to compare with a snow bank camp to boost your incentive to get the work done and get the heck out of there. The first two days my shoulders are so sore I can hardly move at night, but by the end of the week I have delts, lats and forget about a "six pack," my stomach looks like a corduroy road. Every evening the boys bring me rocks they have dug up while planting electrodes (I need 150, one for each hole, and they are hard to find in 30 inches of snow) and they are giving her tarpaper also. By the time I augur my last hole they are right behind me.

The next morning the camp comes down, the X goes out, Bud comes in, and we fly to Kenora. I say goodbye to the Scintrex guy for now. I will see him in Fort Frances in a month. Rod, Ray and I load up the GMC and leave for Sioux Lookout.

Sioux Lookout

The Sioux Lookout job is no big deal. We have four small grids to survey, three by road/skidoo and one by air. We will stay at a motel – no tent this time.

Peter pops in from Thunder Bay and I catch up on some news. KHM is gone. The lease agreement has expired and Noranda has decided to stick to local charter for their flying requirements. Peter tells me I won't be getting my raise, and he puts it this way. "If you get that raise you will be making more money than me". I tell him I have no problem with that, but Peter fails to see the humour intended. Let's face it, Peter can tend to be a tight ass.

Sam is also in and out. He will be bringing in a drill to follow up on our survey work and we reach an agreement. I will go to work for Sam in April with the aim being that I will eventually work into a position as foreman. I'll take a shot at it.

Rod, Ray and I finish the road jobs and in the last week of March we have only the fly-in grid left to do.

We head down to the airbase and climb into a shabby looking 180 with our equipment and snowshoes. Now, I've told you about the hard life these aircraft lead, and this baby has been there-done that for sure. To start with, it is in dire need of a paint job. I know that fresh paint does not make an aircraft fly better, but to me, a neat exterior means someone actually cares about the plane. To top it all off we have a weirdo pilot. Before he climbs in he grabs every control surface and gives them a good shake. First the ailerons, then the tail rudder and elevator tabs. The whole aircraft jiggles and wiggles and the control panel looks like every screw is missing. Behind me, Rod is holding a plastic wall panel in place to keep it from falling into his lap. I'm thinking about bailing out, but the pilot climbs in and I figure what the hey, maybe he is just super careful.

We take off with the tail high, fly 30 miles to our target lake and land with the tail even higher. Holy cow! I have never had this point of view in all my years of flying. I can see the lake less than 10 feet in front of the prop and I grab the sides of my seat with both hands. I'm afraid we will hit a snowdrift and go end over end. Finally, the plane settles on its tail ski and we taxi a mile down the lake to our drop off spot. Now, most pilots will land reasonably close to where you want to head in to the bush, but not this guy. Why he avoids the shoreline is a mystery to me. Slush is a non-issue in late March.

This is a large lake, and he likes to use it all. When he comes in to pick us up he lands so far down the lake that the plane is little more than a speck on the horizon. I wonder if he enters taxi miles in his log book? They would exceed his air miles, I figure.

On the last day there is only a few hours work left on the grid, and Rod and Ray can handle it. I stay in the motel to finish up my last maps for Noranda. I return to the airbase to pick the boys up, and as I am having a chat with the owner, Mr. Scorpion comes in tail high again. I remark on his strange flying habits, and as we leave I see the pilot and his boss having a discussion with a bit of arm waving added.

Sidebar: A year later I will use the same charter outfit to fly out of a bit of a tight spot, and I ask the pilot if Mr. Scorpion is still around. Not anymore he isn't. A couple of weeks after we left Sioux Lookout the guy was at the airport fooling around with the company twin engine Piper. He flipped the wrong switch and retracted the landing gear, dropping the piper on to its belly. As a reward he got to kick his lunch bucket down the road.

Entwine Lake

I have given my notice and Don asks me to do one last job at Entwine Lake. We will be flying out of Fort Frances, so the three of us head south. I will put Ray up in a hotel in town. Rod and I get spend a day or two at home.

We don't have to build a camp this time. Noranda had been at Entwine for a season or two before I started to work for them, and two 12x14 framed tents have been left on site. We presume that they are still standing. Our presumptions are half right.

There will be three or four days of JEM to fill in some gaps. The I.P. guy will fly in to Kenora in a couple of days, and I will pick him up. In the meantime we pick up a grocery order, some stove and lamp gas, and fly in to Entwine with Ray Cameron, my old high school buddy. We arrive at the camp to find one tent standing, and barely standing at that. Two winters of heavy snow have taken a toll on the canvas tents and the one still standing is hanging on by its teeth. Whoever had left this camp two years ago had failed to support the ridgepoles in the centre and had not added additional support under the flys.

The tent still standing has sagged under the snow load and the ridgepole is pretty much U shaped. Both end walls are leaning in at the top, and although the door still opens, the roof is so low we cannot stand upright inside.

The other tent is a dead loss, and it is a sight to behold. It looks like it has taken a direct missile hit. When the snow load got too heavy the tent had virtually exploded. The walls are lying flat on the ground, and a jumble of torn canvas, broken rafters and heavy wet snow covers whatever gear was left in the tent. We can forget about this baby!

It doesn't take too long to shovel the snow off the remaining tent and make it livable. We are lucky that this tent was the kitchen. By nightfall we have salvaged the cots from the other tent and have a nice hot stove to warm us. In early April it is not too cold and with only two weeks of good ice left, the boys are going to have to hustle their butts.

The next morning I get them started on the JEM and fly back to Fort Frances. I will head for Kenora tomorrow to pick up the I.P. guy and more salt. Yes, it's Fife's to the rescue again. I hadn't even attempted to find salt locally. One phone call to Mr. Fife a couple of days ago, and it's no problem – 50 bags of salt will be waiting for me.

I pick up my man and the salt and a day later we fly back to Entwine. I won't be staying - there's no room for me in the one tent still standing, and Rod and Ray know the ropes and the I.P. guy knows the boys. The Norseman stands by while I get the I.P. guy settled in and say my goodbyes. In Rod and Ray's case. it's "So long for now." We will be together again nine months down the road.

Now that I think of it, it is strange that I have never flown much in the Norseman - the workhorse of the north. As we fly back to town at 2000' I ask Ray how far we can glide if the engine quits. He points straight down. Thanks a lot, pal.

I call Don and tell him I am done, Peter can pull the camp out. Don wishes me well and I feel some regret. Don has always treated me fairly. My problems were with the a-hole in Toronto. Ah well, such is life.

I have just a few days off this time. Sam is moving his drill in to one of our grids east of Sioux Lookout and I will be a cross-shift runner. The wife and I use the time off to do some organizing. We have talked this over already. It seems I will always be working near or north of Hwy 17 (Trans Canada) and it makes sense that the family live closer to my work. We will sell our house and move to Vermilion Bay.

My wife drives me to Sioux Lockout and I head in to the drill, which has been moved onto the job already. It's sort of a road access deal, but the last four miles is on a cut, but unimproved right of away, and I ride in on Sam's muskeg tractor. The new road will join Sioux Lookout to Hwy 599 near the proposed Mattabi and Lyons Lake mines. It will serve as a route for Sioux Lookout based mine employees and also as a forestry access road.

We cut three large spruce for our tripod, move the drill into place and collar the hole at 4 pm. The foreman, my helper and I head for the tent to make supper. I will take the first night shift.

The larger drills always run 24/7 with two 2-man runner/helper teams. Twelve volt lights supply illumination in the drill shack and at the tripod platform, and once the hole is started the drill never shuts down until the hole is finished. Shifts run 7 to 7 and the runner never leaves the drill head. The helper keeps the fuel topped up at the drill, pressure pump, supply pump, and the oil stove in the corner of the drill shack, along with other duties too numerous to list. The runner drinks his coffee and eats his lunch with an eye on the water pressure and an ear tuned to the song of the drill. If the bit blocks or the core barrel fills the drill must be shut down before the bit burns in.

The two crews switch shifts only when a new hole is started. Our holes are in the 400' to 500' range, and the final depth is determined by the type of rock encountered. We never shut down in a sulfide zone, and of course, all decisions are in the hands of the on-site geologist. Because road access is fairly good he does not stay in camp. He has a room in Sioux Lookout and comes out every day.

We finish the first hole at 4 am on our third shift, pull the rods and hit the sack for a couple of hours. The geologist comes in to check the core and confirms that the hole is done. We crawl out of bed and help the foreman and cross shift to tear down and move 400' to our second hole. We have the hole collared by 6 pm and the cross shift will run tonight. We get to catch up on some shuteye until 6 am tomorrow.

I'm a guy who likes his sleep and I'm already finding it a pretty tough row to hoe, but I'll stick with it for a while. We will see how things shake out.

Our third hole is back to the end of the well-traveled portion of the road and one mile south, so we cut a trail into our new campsite. Tomorrow we will move our camp and start moving the drill.

We camp on the north side of a small river and our hole will be 300' into the bush on the other side. The river banks are low and the river is only 40' wide at this spot but there is a fair bit of current and the late spring ice doesn't look too healthy. We knock down a few good-sized spruce and build a rudimentary bridge across the ice – just something to spread the load a bit and help the ice do its job. We winch the drill across and are careful to keep the muskeg tractor loads of rods, fuel barrels and pumps etc. light. Our bridge is pretty shaky and we watch it closely.

Before we finish moving the rest of the stuff in, our muskeg tractor breaks down. The differential is cooked. It's no surprise as the diff has been getting growlier by the day, but we had hoped it would last for this hole.

We have a new gear ball on hand and while the rest of the crew snooze I help the foreman pull wrenches. Our tractor is a C10 model, meaning the full width cab is at the front and there are quite a few wrenches to pull. Because the driving axle is at the front we have to remove the cab and cab floor, and with some difficulty we do so. Now we change the gear ball, and I learn a bit of bush-wrench ingenuity.

Every time you install a new differential gear ball, the backlash has to be set. Too much or too little space between the gears will be sure to cause premature failure – and we don't have any fancy feeler gauges.

Sam has been in and out that morning and we ask him about checking the gear settings. "Three and four," says Sam, and here's the 3-4 drill. Using ordinary writing paper you run 3 sheets through the gears. As the teeth mesh the paper should be crimped and not cut. Now you run 4 sheets through, and this time the gears should cut the paper. We give it a try, and bang on – three sheets crimped, 4 sheets cut. We pop that sucker in, replace the cab and the tractor is mobile, with nary a whine or complaint.

Sidebar: Feeler gauges come in many disguises. Back in the day, before computers and electronic spark appeared beneath the hoods of our Detroit Iron, it was common practise to change plugs, condenser and points once or twice a year. Plugs and points need to be properly gapped, the usual gap being twenty-five thousandths of an inch.

The cardboard used in every Canadian cigarette pack is twenty-five thou. If you smoke you have a feeler gauge in your shirt pocket – if you don't smoke – ask someone who does.

You say your manual calls for 30 thou or more? That's a cigarette pack and a little bit. Twenty thou? A pack and a little bit less.

The point being – never say "Whoa," because you don't have the right tools.

While we are repairing Mr. Muskeg Buggy, company has arrived. Another outfit has a drill hole to do on an adjoining claim block and they want to know if they can camp nearby and use our bridge. Of course they can, and while we work, we watch them move their drill in. They load pretty heavy to cross the River Kwai, but our bridge holds.

We have a few problems on our hole. Nothing serious, but it puts us a bit behind and the other guys finish a day ahead of us. We are ready to pull our drill, but since they are already moving out we decide to wait a couple hours until they are finished crossing our bridge. We watch as their last muskeg tractor load appears on the far side, and we know darn well he is not going to make it across.

Their tractor is a J5 model with centre controls. They have removed the little flip-flop cab (you never run with the cab in place in warmer weather), and they have loaded that rascal to the hilt. Drill rods, casing, pumps and hoses, and other drilling paraphernalia are on the side cargo areas. On top of all this they have piled the prefabbed sections of the drill shack. The driver is sitting down in his little cockpit with just enough room to wriggle into his seat.

He pauses for second, starts across, and we hold our breath. He makes it halfway before the bridge collapses and the whole kit and kaboodle sinks in the cold, cold river. It's only four feet deep and not enough to cover the load completely, but it's deep enough to kill the motor and soak the poor guy. He pops out of his burrow, thrashes to shore and we pull him out. He's wet to the skin and is standing there half shocked and shivering. We all laugh like crazy. We are surely a weird bunch.

We hook a cable to their machine and winch it on to dry land with our tractor. While they off load and dry out their engine, we cut more trees and rebuild our bridge. The day is shot by now. We hope the freezing temperature will set the ice up overnight and it does.

At daybreak we carefully pull our drill across, tear down the camp and haul everything out to the road. Our next two holes are 15 miles east at Umphreville but we take the long way around. We can drive in on a turkey track off Hwy 599. Tonight we will stay in Sioux Lookout.

Umphreville in 1970 is a remnant of better days. A few houses, mostly empty, a school with five students (no kidding) and a nearby fly-in fishing lodge make up the whole shebang. It must have been a bustling little burg in the past. Oscar Styffe Lumber of Fort William had a saw mill and camp here in the early years, but had moved on some time ago. All the camp buildings are in pretty good shape. In the old office/commissary are rows of cough syrup on dusty shelves. At least 20 two-gallon glass jugs are lined up waiting to be sold to some poor guy with a bad cough (nudge nudge, wink wink). I dust off a jug to get a closer look, Yup – cough syrup all right – cough syrup containing 11% alcohol. It must have been a busy spot on a Saturday night.

I also find a cancelled cheque written in 1921 on the Bank of Toronto for 2000 bucks. Wowee! Someone had gone out to town pretty stakey, for sure!

There are a few old roads in various states of disrepair, and the only vehicle in town is a beat up 1928-29 model A pickup, belonging to the little fishing lodge. (I returned to Umphreville five years later and bought the little Ford for 100 bucks, passed it on to a friend who sold it to a guy whose garage burned, taking the pickup with it.)

We have two holes here, and we rent a semi-livable old house in town. These will be my last holes for Sam. I've talked it over with him. He already has a good foreman and I don't fit in. I'm a pretty poor excuse as a driller and don't care much for the noise, grease and 12 hour shifts. When we finish the second hole we part on good terms. I am going to hit the contract trail.

Knot Lake

My wife has sold our house, and we move to Vermilion Bay, where we rent a nice little cottage nestled in the pine trees. She goes to work in a restaurant. I get some business cards printed and put out some feelers. I also get to spend some time with my girls and life is pretty good.

Table Scrap

We sold our house to Elmer Corrigan, an old-time prospector in the Rainy River Valley. I wish I had known Elmer better. He was a gold guy, and he was also a master story teller. Here's one of his tales:

It seems that someone had given Elmer some bad firewood advice. Elmer was told that the best all-round wood to burn was standing dry poplar. (Now everyone, including Elmer, knows that burning standing dry poplar is the equivalent of tossing crumpled up newspaper into the stove. But this is Elmer's story, so we will let him tell it.) So in May Elmer walks into the local Legion for a cool one and all the boys remark on the fact that Elmer has not been seen since last fall.

"I've been pretty busy," says Elmer, "The only person I saw all winter was my wife, and that was when we met at the back door, me carrying in an arm load of poplar and her hauling out a pail of ashes."

I get to spend almost a month at home but I'm getting a little nervous. The phone isn't exactly ringing off the wall and I'm beginning to wonder if I've shot myself in the foot again. Then Don calls me from Thunder Bay and asks me if I'll cut 10 miles of line for him at Knot Lake, about 30 miles south-west of Ignace. It's a fly in camp and Don already has a crew on the lake. I can stay in their camp, eat their grub, and they will even supply a helper to assist in chaining the lines. I'll make $900 for two weeks' work and it is a sweetheart of a deal, (Looking back through the haze of 40-some years I now know that Sam had given Don a nudge.) My wife drives me to Ignace and I fly to Knot Lake to start my contacting career.

They have three tents on the lake and I get to bunk with Ozzie – the same Ozzie who went on an enforced diet north of Nakina a few years before. Ozzie is doing some prospecting for Don on this claim block and a couple of my former Noranda mates are doing some geophysics on a few miles of existing grid. I am going to extend the grid.

Now, Ozzie is a nice little guy, but it doesn't take long for him to start driving me a bit nuts. For one thing, Ozzie is a snorer. He snorts when he's awake and he snorts in his sleep. He must have trouble with his schnozz I figure, but waiting for the next snort is like waiting for the other shoe to drop, and I sleep very lightly. Ozzie is also a frustrated Madison Avenue advertising layout guy. He has a stack of his ideas and shows me every one, over and over. He is actually a pretty good cartoonist, and one ad idea is for insect spray. A mean looking mosquito is being spritzed by a can of spray with the logo "Bug R Off" on the label. It's sort of cool, but when Ozzie shows it to me for the tenth time I kind of wish I had a can of "Bug R Off" to use on him.

The worst thing by far is that his mom has given him a book to read at camp. It is the Guinness Book of World Records, and Ozzie is determined to share the whole darn book. It's "Did you know?" and "Guess how many," on and on, over and over!

One evening I am having coffee with the two Noranda guys under the patio fly. Ozzie is in the tent and he pipes up, "Bet you don't know how high the tallest tree in the world is."

"No, we don't."

"Take a guess," says Ozzie.

Someone had brought a newspaper in to camp a few days ago and I had seen that same tree info in a "You Ask Us," column. "Keep him busy boys," I say, and I'm hunting for that newspaper.

So the guys are making random guesses and Ozzie is having a great time saying "No," and "Not even close," and giggling away.

I find the paper. "Is it 328 feet 5 and ¾ inches?"

Dead silence – followed by a little voice. "How the heck do you know that?" We all laugh like crazy, and we never hear from Mr. Guinness again.

A couple of nights later I am dozing off, when Ozzie gives a loud snort. "Wake up Ozzie!" (I'm a little tired of having my shuteye interrupted.) Ozzie sits up looking bewildered.

"You're snorting again," says I, and just then there is another loud snort, but it's not Ozzie! It's coming from outside the tent wall right beside my bed. The other two guys always stay up later than Ozzie and me, and I figure they are having a little fun.

It's mid-summer and it's still half-light, so I whip outside to give them a piece of my mind. As I round the corner of the tent I come face to face with a big black bear eating our Oreos! It's hard to say which one of us is more surprised, and we both back off so fast we go ass over tea kettle! I go back in the tent and tell Ozzie we have a bear and I have chased him away. Maybe he won't come back – like hell he won't. Fifteen minutes later the bear is looking for more cookies.

The sun has set by now and darkness is gathering, so we light our Coleman lamp and hang it in the patio. As usual, in the summer months, we keep our bread and cookies in a steel garbage can with a lid. This is to keep the mice and creepy crawlers out, but it's no deterrent to a hungry bear. I move the can to the lit area in front of the tent, but all this does is bring the bear closer. As far as he/she is concerned those are his/her cookies. I peg a few rocks at the intruder but am soon out of ammo.

By now the ruckus has awakened the two guys in the next tent and they come over to see if there is a party happening. As we tell them what's up, another bear shows up at their tent. We realize now that trying to get any sleep is futile, so I put on a pot of coffee.

By midnight we have four bears in camp – one behind us, one south of the guys' tent, one to the north, and one between us and the lake. Our tents are on high ground 75' back from the shoreline, and we can hear the bear in the willows along the shore although we can't see him.

One of the Noranda men knows his dynamite, and there is dynamite in camp to use for trenching. He short fuses a stick and throws it at a bear. It doesn't hurt the bear any, but the loud explosion chases them away only to return a half hour later. The night goes on. Drink some coffee, throw some dynamite – repeat.

At 4 am the darkness starts to give way and by 5 am we can see our enemy. North and east of us are two good sized black bears. The guy on the south is quite small, probably a yearling. We can see him coming up the rocky shore line, so dynamite man short fuses another stick and gives it a good toss. It goes over the bear's head and lands in shallow water. The little guy heads out to see if he can catch that fish, and when he gets within five feet the dynamite goes off, shooting a plume of water 50 feet into the air! The yearling does a double back flip and the last thing we see of him is his bum as he hits the trail south at 100 bear-miles per hour. The two big blacks ones have had enough also, but not the one between us and the shore – this guy is determined to stick around.

Dynamite man has had enough. He's going to shoot that bugger, and he gets his gun from the bottom of his packsack. Now, our bear guns are usually of the 303 British Enfield variety, $19.95 at your local army surplus – but not dynamite man. He has a 7 mm 9-shot Beretta pistol. It's definitely not legal in Canada, and I'm a little worried that a calibre of that size will be more likely to piss the bear off instead of killing him.

However, daylight is coming and when we can see the bear outlined against the back drop of the lake, dynamite guy empties the clip into him at a distance of 50 feet. Silence. We wait half an hour and very, very carefully investigate, and find one dead as a doornail, cinnamon colored bear. We feel kind of bad about it. We wish he had taken the hint.

The Dept. of Lands and Forests has to be notified, of course. We don't go to work today. We have some sleep to catch up on, and when the conservation officer comes in that afternoon, he listens to our story and all is ok with him. He doesn't even ask us what was used to kill the bear – don't ask, don't tell. He also tells us that the brown colour is rare among black bears, but the colour mutation is always accompanied by a very aggressive bear personality. He understands that the bear had to be shot.

We pull the bear to the water and tow the carcass down to a creek at the far end of the lake. Buzzards and eagles will perform the last rites.

103

Of course our evening fact-trading session at Coffeepot Corner now deals with stuff we know about bears – all absolute truths: (Keep a salt-shaker handy.)

Fact: Bears have poor eyesight. You may think they are attacking, but they are really just trying to escape. (One grain of salt gone.)

Fact: Bears respect size. Stand tall with your arms outspread, and holler loudly. (Ok, I'll buy that, mostly because it puts me in a good position to bolt for the nearest tree. But wait – if he is myopic, he would have to be pretty close to tell how tall I am – another grain of salt gone.)

Fact: If you climb a tree, pick a skinny one. It will be more difficult for the bear to climb. (Ok, size matters – worth a try.)

Fact: Bears don't like to be hit in the mouth. (Sounds good – I don't like it either, and if I'm up the tree ahead of the bear, I'll get a couple of good kicks in before he eats my boot.)

Fact: If all else fails, assume the fetal position and play dead. (No one likes this one – we all choose to go down fighting.)

Fact: Bears are as much afraid of you as you are of them. (But, have you ever heard of a bear going back to his den to change his underwear?)

We are having a great time, and each "absolutely true bear fact" is greeted by more laughter than the last, and our tummies are getting sore. Ozzie, however, is uncharacteristically quiet. Maybe it's the size thing – Ozzie couldn't scare a bunny rabbit.

Bears on the picket line don't cut much ice with us. We are young, will live forever, and we always have the great equalizer, our axe, in hand – but Ozzie starts to find excuses to stay in camp. He wants the Beretta, and dynamite guy reluctantly hands it over.

We return one evening to find an excited Ozzie dancing around. A bear had visited the kitchen, and Ozzie had touched off a cap in its direction. He was too nervous to do a follow-up inspection, so we check it out. No sign of a bear, but the spaghetti pot has taken a direct hit.

Those Noranda boys love their pasta. Ozzie is less popular than the bear until the spaghetti pot is replaced, and from then on, dynamite guy guards Ozzie at the showing.

Table Scrap

I'm not saying that Ozzie was trigger-happy, let's just say he was a tad nervous. However, a twichy finger can lead to stuff happening. As a case in point, read on: (This was actually in a farm news publication in the late forties. Even then, as an eight-or nine-year-old, I laughed like crazy when I read it.)

Johhny was two hours late for school. The school-marm wanted an explanation or the strap would come out.

"Ma'am," says Johnny, "Last night a fox got into our chicken coop. Old Duke raised the alarm, my dad jumped out of bed, grabbed the double-barrel twelve gauge and went to investigate. Now, my dad sleeps in the raw, and he was in a hurry. He cracked the door to the hen-house, lined up on the fox and old Duke cold-nosed him from behind."

"Teacher, ma'am, we've been plucking chickens since four o'clock this morning."

Wе don't have any more night visitors for the next two weeks, but from time to time we return to camp to find evidence of another bear shopping trip. Someone has heard that coating a can of Raid with honey will entice a bear to chew the can, so we try it. That afternoon we return to camp and find an empty can of Raid punctured by bear teeth. I can imagine that bear shaking his head saying "P'tooey, p'tooey, p'tooey!"

(I will leave camp 10 or 12 days later, but I heard that the bears returned, forcing the boys to move to an island. The bears tracked them down again within three weeks, but by then the job was finished.)

Two days after the bear incident my line cutting is interrupted. A guy in Sioux Lookout has heard that Noranda is on a hot property and the word is out that he is bringing in a staking crew. Don wants to protect the showing, so he brings in the contract stakers. Or, to be more exact, he brings in one staker. Bill and Dave usually do Don's staking, but Dave is busy elsewhere. I will stake with Bill and share the contract.

We have 90 claims to stake and we get right at it. Ten hours a day and no lollygagging, and its hot – boy, is it ever hot – not a cloud in the sky and 90+ degrees every day. By 10 am my clothes are so wet I might as well be working in a rainstorm. I can't carry enough water or pop to stay hydrated, so I drink at every pond or creek I come across. A large section of the claim group covers a flattish out-crop area with lots of blueberry bushes and the blueberry crop is stupendous this summer. As I trot along I scoop up handfuls of berries without slowing down. I wonder why those dumb bears prefer cookies.

This is bear country for sure. Bill tells me one evening that a bear had kept pace beside him all day, no doubt unwilling to share his berry patch with a human.

One of our return trails to camp crosses a shear zone and nestled at the bottom of a 20' rock wall is a little pool of water at the mouth of a tiny cave. Ice from last winter must still be at the back of the cave, because the water is clear and oh, so cold. What a find!

Dynamite guy tells me that he can't count the times when a waitress has served ice water with a meal and he doesn't drink it. He says he will never ignore ice water again.

Ten days later, the 90 claims are finished and I am done line cutting. Don now has 90 + miles to cut, and will bring in the pros from Val D'Or. I head home, 30 pounds lighter and with 5000 bucks in my pocket. WOW! Almost a year's wages for less than one month of work – I'm hooked.

The wife needs reliable transportation, so we go to Winnipeg, lay out 2500 bucks cash for a two-year-old Impala hardtop with 30,000 miles on the clock. and we are still pretty stakey. Our trusty old Pontiac will become my bush car.

Umex Rush

Once again it's a waiting game. I'm doing some telephone schmoozing but I am starting to think that my trap line may be too small. September passes, and by late October the old bank account is showing some wear and tear.

On October 28th the phone rings. Union Minière of Belgium (hereafter referred to as Umex) has found a zinc deposit next door to the old Central Patricia gold mine at Pickle Lake Ontario. Les C. wants 100 claims staked at Pickle Lake PDQ. Whoopee! Super Staker swings into action.

I have a tent and camp gear. What I don't have is stakers, or much cash. I am so green I didn't ask for up-front money, for crying out loud! Too late now – I'm to meet Les in Pickle Lake in three days, and I'll have to foot the bill for a while.

I keep the phone hot for a few hours and come up with a basic three-man crew, Dode, Donald and Ed. Dode is an experienced staker from the Rainy River Valley and will drive up tomorrow morning. Donald is a local guy and an unknown quantity, although he talks a good game. He will supply his own transportation and will join us in three days. Ed is an older local fellow I first met when working the Bridges Township property in 1968. I don't expect Ed to do any staking – I figure he will be helpful in a non-staking role and I'm right. Ed will prove to be indispensable.

The next morning we load Ed's pickup to the hilt, I give him 200 bucks for walking-around money and he heads for Pickle. He takes the camp gear, a rented Skidoo and a few groceries, and I tell him to find a room in Pickle for the night. Dode and I will follow later on today in Dode's car. I leave the Pontiac at home – we won't need a fleet of vehicles.

Dode picks me up before lunch and we head for Ignace where we will turn north off the Trans-Canada for the 200 mile drive to Pickle Lake. At Dinorwic we take a 35 mile detour to Sioux Lookout where I visit the recording office to buy 100 sets of tags and some claim maps. I plan to stake the claims on my licence.

By the time we reach Ignace and turn on to 599 it is starting to get dark and when we pass the south end of Sturgeon Lake it's snowing pretty heavily – the first snowfall of the year. The snow starts to build up and we see some signs of grader activity, but we can't quite figure out what the guy is doing. It appears he's hop-scotching, plowing a couple of miles and then running with the blade up. About 30 miles from Savant Lake we catch some deeper snow on a corner and spin out into the ditch, probably the only ditch in the 200 mile road devoid of rocky outcrop. We get back on track easily enough and I tell Dode not to push it. We will get a room in Savant Lake and continue on to Pickle Lake tomorrow. Dode says he is going to ask the first drunk he sees if he is the grader operator.

We check in at the Savant Lake Hotel, eat supper and hit the bar. Sure as shooting, there is a guy at a table obviously half in the bag already, and Dode asks if he is the grader guy.

"Dam straight" he says, and knocks back another glass of bubbly. Dode asks why he didn't plow the whole road while he was at it.

"I was in a hurry" the guy says, and orders another beer. I'm starting to love this place already.

The next morning we hit the road early. The snowstorm has blown through and the road is pretty clear. We arrive in Pickle and run down Ed at The Landing. He had scored us beds at a boarding house and there I get to meet Les and his sidekick Louis.

Les has his own maps, so we spread them out and plan our attack. The Umex find is just north of the Kawinnigans (Crow) River and not too far west of Hwy 599. Les wants to tie on some claims in Ponsford Twp. north and west of the Umex group, so we head out to set up the tent and start staking.

The bush looks pretty good. The terrain is flat and in years past had been stripped of trees for mining timbers. Most of the bush is 20-foot high second growth jack pines and the whole area is criss-crossed with trails. We drive in with the pickup a couple of miles, find a claim line and unload. Ed and Dode start to set up the camp and I take the Skidoo to find our starting spot. There is only about 3 inches of snow accumulation, enough for the Skidoo but not enough to deter the little Ford half ton.

It doesn't take me long to find out we are in enemy territory, so I head back and tell the guys to hold it, we will have to move farther in. This time I tell Ed to wait while Dode and I will do some exploring, and so we continue on by Skidoo.

We run across a couple of Umex stakers and they tell us that the next four or five miles of ground has already been grabbed. They are just doing some checking, crossing the I's and dotting the T's, so to speak, so it's back to the boarding house with the bad news. I point out that even if we could leapfrog the other stakers it would be pointless. We would be outside the greenstone belt and into granite – just worthless moose pasture.

Les is undeterred and produces a set of airborne magnetic survey maps (government maps available at any district geologist's office.) We will stake mag highs, and we circle areas of interest on the maps. Three targets are not far off 599 and are no problem – we can access them from the highway. Another target is seven miles east of 599 and two miles south of the old Pickle Crow mine. This may pose a problem, as there are no old trails on the east side of the highway. The seven-mile road into Pickle Crow is no longer maintained and a large swamp between that road and the proposed claim block will be impassable until we get a good hard frost.

Now get this: the three highway claim blocks we outline will add up to 189 claims. The Pickle Crow block will be around 100 to 110 claims, and they want to pick up more ground 20 or 30 miles west of Pickle Lake. It's plain to see I've been a victim of a bait and switch ploy, but what the heck – never say "Whoa" in a snowbank.

First of all, I need some serious up-front cash. My bank account is dwindling and I squeeze 3000 bucks out a reluctant Les. Tough bananas Les, either cough up or we hit the road. Les coughs it up.

Donald arrives and he and Dode will start the first Hwy. 599 block tomorrow. Ed will drop them off in the morning and pick them up at night. I decide we will stay in the boarding house for now. I will work the phones - I need more stakers and I need them right now!

Pickle Lake/Central Pat

It's time for a little bit of back-ground. There are two townsites in the area, three if you count the abandoned town of Pickle Crow. Central Pat is just off Hwy 599 where the 5-mile road leading to Pickle Lake branches off, and is not much more than a memory of days gone by. It still has a good hotel/bar, a post office and an Ontario Provincial Police detachment (O.P.P.) A few people still live here and a few empty buildings in various states of disrepair still exist.

Five miles west is the town of Pickle Lake (aka The Landing) and this is the main business/residential area in the district. It also has a hotel/bar and the main street runs along the shore of Pickle Lake itself. South of the hotel we find a general store with groceries, hardware and lumber supplies. A post office and a liquor store are mixed in with some neat houses.

A bit further south is a small complex of trailers where an old Inco acquaintance lives. Bert is an area supervisor and he will be putting a crew into the bush north of Pickle this winter.

107

Another couple of miles farther south is the Pickle Lake airbase, a gravel strip with a few planes sitting around. A more or less daily sked to Sioux Lookout flies out of this strip and in the summer months it serves as a destination for sports fisherman arriving from south of the border.

Just north of the hotel is the Hooker Air seaplane base and on the beach sits a Bellanca Airbus on floats waiting for the spring thaw. The Airbus is a huge single-engine workhorse from the thirties, and this one may be the last survivor.

Sidebar: George Fournier from Lac du Bonnet, Manitoba has looked after the Airbus for years, and still comes up to Pickle to do the annual Certification of Airworthiness on the Bellanca. It was George's step-son who ferried Carl, Norm and me in and out of Bissett in 1963. (The Olympic Rings interlock again.)

Owned by Superior Air Services of Thunder Bay, it sat on the beach near the Hooker Air office awaiting spring break-up.

Just across the road from the hotel is the restaurant/rooming house where we are hanging our hat at the moment, and around the corner is Koval Brothers Transport, the main supplier of freight movement for the north. They also have a terminal in Winnipeg.

The road into The Landing joins the main street right beside our boarding house, and the empty lot on the opposite corner belongs to Moose. Moose is a big old grey mare at least 30 years old and is the last survivor of the days when trees where hauled out to a long-gone sawmill to become mining timbers. She had a tough life, I'm told, but now she enjoys her retirement, spending her days munching on goodies supplied by generous town folk. Koval Bros. brings in timothy hay for her, but Moose prefers lettuce and carrots which she shares with her posse of ravens. There are always ten or more of the huge birds hanging around, with some sitting on her rump as they wait for the next grub delivery. The Landing has only one rule. You don't mess with Moose. Pretty cool huh?

Old Moose knows a thing or two. Just across the road the hotel has a short stairway leading down to a basement storage area, and she knows that lettuce and stuff appears from this spot from time to time. One day the food delivery was a bit late, so shee decided to help herself. She went down the stairway and got stuck. She couldn't back out, and she couldn't turn around. An SOS went out, and soon there was a whole slew of people around, each with a different solution to the problem. Finally, a sling of sorts was fabricated and Koval's big front-end loader picked Moose up and set her on level ground. Both Moose and Pickle Lake learned a lesson that day. She returned to her posse and the goodies arrived on time from then on.

One evening one of my stakers is returning from the bar when he decides to take a horseback ride. He grabs a handful of Moose's mane, but before he can swing aboard she bites him on the ass – not just a nibble – a darn good bite. Another lesson learned. Moose can take care of herself.

Making one phone call is no simple task, and working the phones to line up stakers is almost impossible, but I persevere.

I have only 3 phones at my disposal. There is a pay phone in each hotel lobby and a phone in the boarding house landlady's bedroom. Calling cards are not in use yet, and I will have to charge all calls to my home phone, so I give my wife a heads up and ask her to make sure the phone bill is paid. I go to the Central Pat Hotel and find out that the pay phone won't work. The town is now crawling with stakers and the coin box is full to the hilt. I can't even feed a quarter in to call the operator, so I go the hotel at The Landing. The coin box on this payphone is not yet full, and I soon find out why. There is a broken wire in the line leading from the hand set to the phone, and although it works some of the time, I stand a pretty good chance of an accidental disconnection.

Using the landlady's phone has its own perils. She readily admits to being a periodic alcoholic (as she puts it) and has found a willing drinking partner in Les' assistant Louis. She keeps the phone in her bedroom, supposedly for security reasons, and I find that I have to choose my phone times pretty carefully. If she is any more than three sheets to the wind, she considers any man in her bedroom fair game, if you get my drift, and it's hard to concentrate when you are fending off a horny old lady.

So I learn how to use the broken wire phone and it is a complicated operation. There is no furniture in the hotel lobby and the payphone hangs on the wall. I have two sheets of paper, one with a list of contacts and the other for jotting down notes, and I stick them on the wall. Then I find the right position to keep just enough tension on the wire to maintain a steady dial tone and still be close enough to the phone to reach the dial and coin slots. (I have a supply of coins lined up on top of the phone).

Now I start to make my calls, standing very, very still. If I cough, burp or fart I will break the connection and have to start over. From time to time someone else wants to make a call, and I reluctantly relinquish the phone. They soon give up in disgust and I'm sure not going to offer any advice. Let them figure it out for themselves.

I call Bell Canada's service department to report the broken phone and they tell me the service guy will be in Pickle in two weeks on his monthly tour. "Any chance of a special trip?" "Sorry sir, no exceptions." The service guy comes in two weeks later and repairs the cord. Does he empty the coin boxes? No he doesn't. That job belongs to the coin box guy and he comes in next week.

This blows me away. They come 300 miles from Dryden to do one job? Ma Bell's idea of workplace efficiency, I guess.

But I'm making some headway now. Ray and Tim are taking a break from Noranda and they are on their way. I work my way through the grape vine and hire Art, Victor, Rosaire and Jerry from Savant Lake/Sioux Lookout. They all know each other and have done some staking in the past. They will work together as a four-man team.

Another complication! Les has people to do and things to see. He is leaving Louis in charge - and Louis is drunk every day. He can barely look after himself. Les hits the road south never to be seen again. I'll have to see how this plays out.

Ray and Tim arrive, and I put them in with Donald and Dode. Good old calm and collected Ed is doing his thing running men up and down 599, and I lean on him a lot. Art and his crew arrive and I give them a block east of the highway. These are good men and cause me no trouble at all.

Now I have eight men staying at the boarding house, not counting Ed and me, and it is getting far too expensive. A camp is out of the question. I would need three tents, lumber, and a place to set up. No way, Jose.

It's also getting tough to feed the men. The landlady is not even pretending to run the restaurant. Sometimes she will cook a roast or a pot of spaghetti and it's every man for himself – just dig in. The stakers are working hard and need three squares a day, so I make a deal with the hotel across the road. She reluctantly agrees to feed us at a second sitting, but after two days her staff complains, so she shuts us down. No problem, Plan B is just around the corner.

I talk to Bob Parker at the Central Pat Hotel. He can't help me out. His cook is sick and he and his wife have their hands full. Their rooms are all rented, he runs the bar 12 hours a day, and she cooks only for their live-in customers. Bob suggests I rent the old Canadian Legion in Central Pat, and he gives me a phone number.

It turns out that the Legion is now owned by a guy from Osnaburgh, a native community eight miles south. He will be glad to rent it to me, and trots into town to show me the place. It is an old log building about 24' x 40' and is in good shape. Just one large room and a kitchen, but the electricity is on, the oil tank is full and there is a propane cook stove. The water pipes are frozen but that poses no problem, there is an outdoor biffy behind the Legion. I have brought lots of kitchen stuff with me, far more than I needed for a three-man camp, and it sure comes in handy. I borrow a dozen steel cots and mattresses from Bert (my Inco buddy) and Ed swings in to action. In one day he has the beds set up dormitory style, stocks the kitchen, picks up the men and cooks supper. Ed has also rounded up some water containers, which we fill at Bob's hotel. That evening we move the men into the Legion. Ed and I keep our rooms at the boarding house. Ed works 12 hours a day and needs his sleep, and I need my room for office space.

I have to let Donald go. I have tried to use him as a foreman, but the men don't like him. Ray tells me he is lazy. Donald just wants to chase men around without doing his own share.

Truth be told, I don't like Donald too much either, and we part on bad terms. I give him a bit of a bonus though. After all he was willing to come to Pickle when I needed him.

Dode is leaving also. He wants to take some time off before his own winter Inco season starts, and besides, he thinks I'm a little nutsy. He is probably right. I'm getting frustrated with Louis and the money situation. I have to be careful. Every time a man leaves I have to pay him, and I'm almost broke. I manage to get another thousand from Louis, but it's just a drop in the bucket. I already owe in the neighbourhood of 2000 bucks in day wages and a lot of things require a cash outlay. Only Ed knows the dire situation I am in. In addition to all his other duties, he is my whining post. I don't tell the other men, they would surely bolt if they knew.

We need a chopper. There is no way to get to the western claim blocks or the block south of Pickle Crow without a helicopter. There are no lakes in either area and the local air service won't put a plane on the ice yet anyway. Snow cover and lack of really cold weather has retarded ice build-up on Pickle Lake. I try to convince Louis, but he tells me Montreal won't spring for a chopper. The ferry costs alone from Ottawa to Pickle Lake are far too expensive.

I'm almost at the end of my tether. I've been to Sioux Lookout (by sked) to record claims and buy another 300 sets of claim tags, all out of my own pocket, which by now contains little more than lint. My stomach is sore most of the time, my left eye has developed a permanent twitch, and I'm living on coffee and cigarettes. It's make-or-break time I figure, so I find a crumpled up ten-dollar bill in my jacket, head to the liquor store and buy a 2-4.

I corner Louis at the boarding house, plunk the beer down between us and start feeding Louis some suds. Louis is going to tell me a story, and it better be a good one. I've had it up to here.

It takes a few beers to prime Louis' pump, but finally the story starts to emerge. It turns out that a Montreal syndicate had given Les a contract to stake 500 claims at 100 dollars each. Les had brought Louis in as a minor partner, and needed a simple-minded bushwhacker to do the staking. Les chose well – I am as simple as they come. He left Louis with some of the cash to run the contract, and took the rest of the 50 grand to Dallas to play the silver futures with the Hunt brothers. (Lamar and Bunker Hunt were on a mission to corner the silver market.) Les will return with a wheelbarrow full of money and we will all get a bonus. I tell Louis, "Good luck with that. Now give me a name and number in Montreal."

Louis doesn't want to part with any more info, so I show him the hammer. "Keep talking, my friend." By now tears are rolling down his cheeks and he wants to know what will happen to him if he loses control of the job. Though titty miss kitty, I want that number and I want it now! Louis coughs it up and I leave him to soak up the rest of the beer. I have a phone call to make.

I call Montreal, get hold of a guy I'll call Abe, and tell him the whole story. Abe is astounded. He is unaware that Les has absconded with most of the money, and is also unaware of my request for a chopper. We have a long heart-to-heart, and Abe sounds like a guy who knows what the bear did in the buckwheat. Certain revisions are needed vis-a-vis my contract, and Abe is cool with my demands.

First of all, I need some money. I'm more than five grand behind by now, and Abe says "Fine." I will get my 5000 bucks forthwith, and more when I need it.

Whew! One rock out of my packsack. Abe tells me I'll get my chopper, and I throw another rock away.

I have one more demand: I'm willing to stick to my original agreement on the first 500 claims, but if more than 500 are to be staked, I want the same deal he gave Les – 100 bucks a claim, and I do the recording – Abe agrees.

We luck out on the chopper. Abe calls Viking helicopters in Ottawa and they have a Bell G4A being ferried from Yellowknife to Ottawa. The pilot, engineer and all their equipment and gear are aboard, and they are cool with a two or three week stopover at Pickle. An extra three weeks' wages will come in handy at Christmas.

Two days later I meet them at Hooker's airbase and show them the setup at Central Pat. An empty lot beside the Legion will serve as an excellent landing pad, and a street leading south to the Central Pat Hotel two blocks away will be perfect for lift off and landing. Since we are alone at this end of the village, security and road traffic will be no problem.

The rush is starting to settle down a bit, so I score a room and meals at the Central Pat for the chopper guys. They can walk back and forth to the chopper and will fuel up at the airbase. All the associated expense involved is taken care of by Viking - I don't have to finance any part of the chopper deal. Due to the Les debacle I am still a little gun-shy regarding finances, but the fact that Viking is obviously being taken care of makes me feel a bit better.

I'm not out of the woods yet though, not by a long shot. I'm still spending money I haven't got and my house of cards is as shaky as ever.

Stakers come and stakers go. In anticipation of the chopper arrival I have called Bill in Thunder Bay. (Bill, you will recall, was my staking partner at Knot Lake.) Bill and his partner Dave arrive with four men and their camp gear the same day as the chopper comes in. They will head west tomorrow by chopper and start to stake three claim blocks totalling 270 claims. They will sub-contract at $50 per claim and supply their own grub. I will write one check instead of twenty.

These guys are professional prospectors and carry a pretty good reputation in the Thunder Bay mining exploration community and I don't want to jerk them around. Before they head in I lay it on the line. They will get paid when I get paid. Will they trust me? Yes they will - cool - I don't even trust me.

Bill and Dave are no trouble at all, at all. I give them claim maps with the area to be staked outlined and they handle the rest. They have their own tags and recording forms and set up their own schedule with the chopper pilot regarding camp moves and grub runs. When the job is done I am handed the completed recording forms and claim transfer documents. Pretty cool, I'm thinking. I did make one trip into see them, but we had to fly around a bit to find anybody. They are in the bush from dawn to dusk. I just wanted a chopper trip anyway.

Art and his crew are no trouble either. They plan their own daily work and draw up their own recording forms. I oversee their work and Ed feeds them and shuttles them. Their roadwork is almost done and they will chopper in to a 60 claim block five miles off Hwy. 599 in a day or so. No problem.

Ray requires a little more attention than the others. He is a good man but lacks the experience of the other crews. I work with him every evening to plan the next day and do the recording forms for him. Also, Ray's crew is in a constant state of flux. He still has Tim with him, but I am putting Ray into the 99 claim block seven miles east of 599 and two miles south of the old Pickle Crow mine site. It's a big job for two men and I try to find some reliable assistance.

Bert tries to help me out. He has pulled two men out of his camp north of Pickle. They are heading back to Fort Frances for their Christmas break and Bert asks them if they want a few days extra work. They agree to help out, so they join us at the Legion hall.

One could record only 18 claims per application. With 728 claims to record, and keeping in mind that not every block was a perfect rectangle, we would have used at least 60 forms.

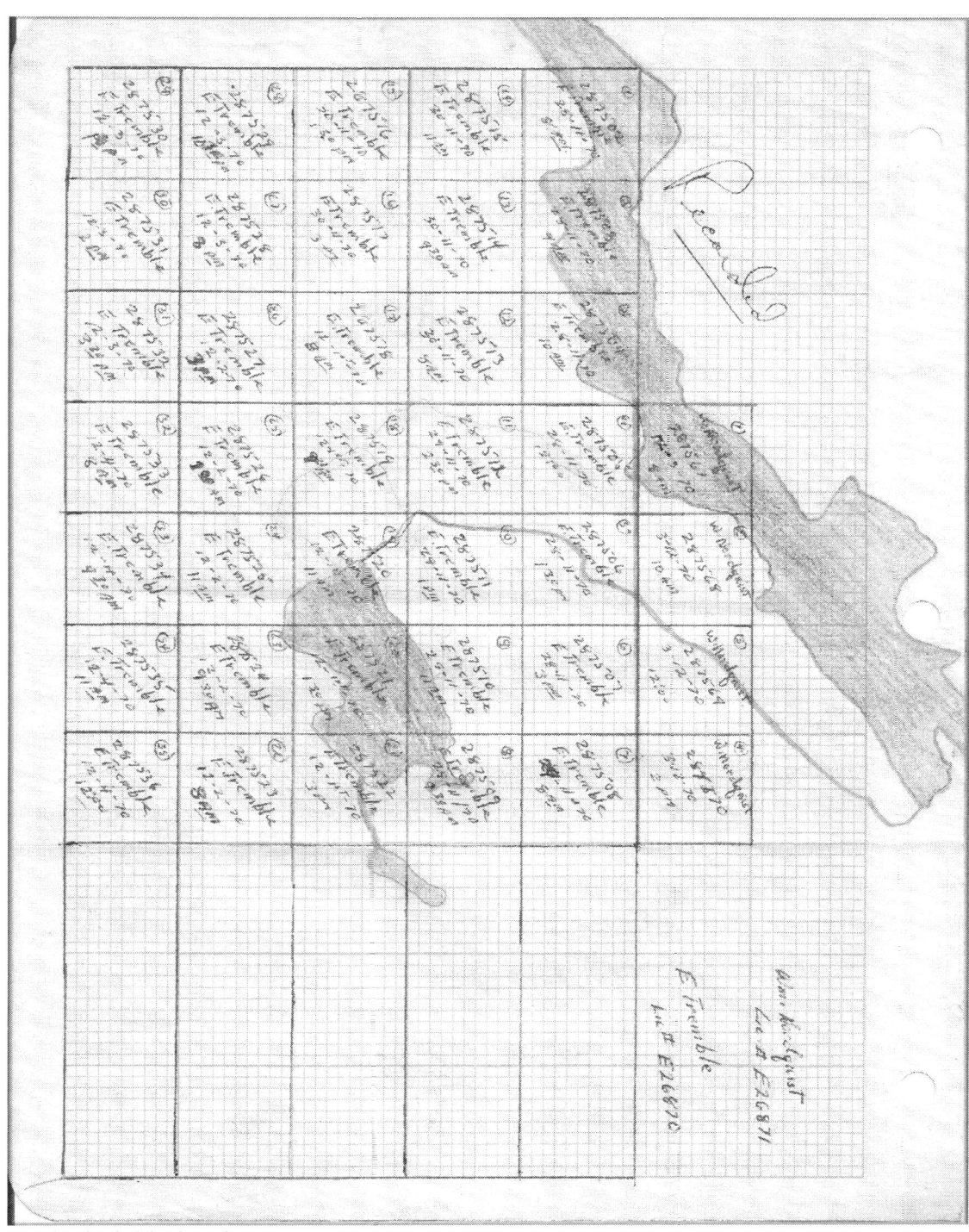

Typical Field Sketch

William Nordquist's licence wou_d have been used for the 15 claims to be staked the next day.

114

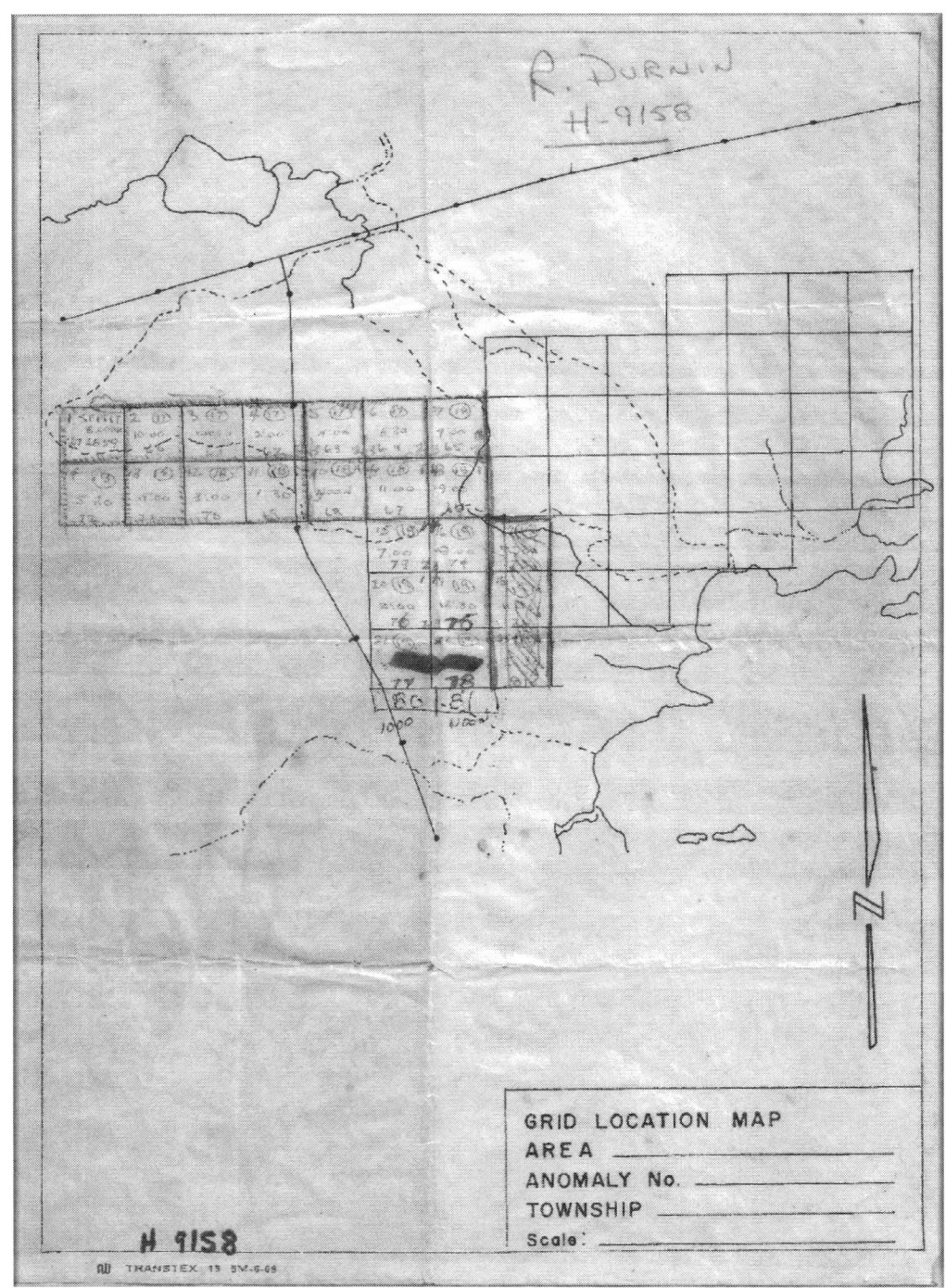

Another Typical Field Sketch

They put in one day with Ray. I go to the hall that evening to check on the next day with Ray and find a fight brewing. Someone has brought in a 2-4 and Ray and one of the Inco guys are tossing a knife at each other. No kidding! They are sitting on the edge of their cots about five feet apart throwing a hunting knife at each other's feet. One guy tosses the knife and it sticks in the floor, then the other guy picks it up and tosses it back. Each time it gets closer to stocking covered toes, and sooner or later someone is going to get stuck. The rest of the crew is pretty quiet, and I can tell they are a little uncomfortable with the threat of impending violence. I step in, take the knife away and tell them for Chrissake go to bed! I hang around a bit and things seem to settle down somewhat, so I go back to my room at the boarding house but I don't sleep too well that night, wondering if I will find dead bodies at the Legion tomorrow morning. I boogie up there early the next day and find a somewhat sheepish Ray, and no Inco guys. They have headed south and I'm not sorry they are gone. I have a heart to heart with Ray and put a ban on any beer in camp in the future. If they want a cool one after supper they can walk the two blocks to the Central Pat, for crying out loud. Ray agrees, but he and Tim will have to work alone today.

Sidebar: One of those Fort Frances boys was a loose cannon, and Bert told me a story I wish he had shared with me a few days earlier. Over a year ago, Inco had been using the Landing hotel as a way station for Bert's men as they came in and out of camp. Mr. Loose Cannon got drunk and obstreperous one evening and the bartender cut him off. L.C. wasn't too happy and he headed to his room telling everyone he was going to trash the joint. Turned out that he had a power saw up there, and soon the chain saw was buzzing and the hotel owner was hammering on the door to L.C.'s room and screaming at him. Loose Cannon never did cut up any important stuff, but as a result of his shenanigans Inco was kicked out of the hotel forever more. (Thus the mobile home complex.)

Table Scrap: LC and the Challenger

I heard about this one a few years later. The time frame was summer '68, probably after I had stopped at Nestor Falls with Rod and Lorne on our way to Red Lake.

That Nestor Falls Inco crew partied hearty. One such party wound down shortly after dusk. I'd assume it was a rain day and the party had started when the sun passed over the yard-arm.

LC had a Dodge Challenger, just out of the show room. With a 427 hemi, six-pack, 4-speed and posi, it was a typical muscle car of the time. He decided to take an evening drive – too many bubblies – no common sense.

They were staying at a tourist camp near a lake. Their cabin was at the foot of the hill.

LC selected 4th gear instead of 1st. The clutch, of course, could not handle all the horses from a standstill. He made it halfway up the hill and the pressure plate blew out of the bell housing, cutting the brake lines and steering shaft. Pieces shredded both fenders in front of the door posts. How LC's legs escaped damage I do not know. One pie-shaped part arced through the air and came down through the roof of a cabin rented by an American family. The lady thought it was a meteor strike. She scooped it up with a dustpan and it melted the plastic!

The car, with no brakes and no steering, rolled backwards down the steep hill until a big pine stopped it, resulting in the back bumper and trunk lid ending up in the rear seat.

The car was a total loss – Mopar replaced it on warranty.
Believe it or not – it's in the book!

A local guy comes by with his son to see me. The lad wants a job and would I take him on? He is a big, strong looking, polite young fellow and doesn't say much. I wonder why Dad is doing all the talking, but what the heck. Maybe the guy is just shy and probably overawed in the presence of Super-Staker. He has never staked a claim, but Dad says he knows his way around the bush and which end of the axe to grab. I like a guy who shows initiative, and every staker has to spend day one in the bush sometime, so I say, "Sure, be here before 8 am. The chopper leaves at first light." Before they leave I give him one final heads up. He will be working with Ray, and he will do what Ray tells him. The kid nods – he's got the picture (or has he?)

Let me tell you about our chopper pilot: Ken Venturi is from Massey, a small town near Sudbury. I doubt if he has celebrated his 25th birthday yet, and I'm not sure if the ink on his ticket has had a chance to dry. No matter, I can tell from the start that he is a good pilot. He handles the chopper smoothly and competently, and everyone likes him. The boys tell me that he always finds a spot near their daily starting point and never complains, never. The best sign of a good pilot? His engineer likes him and engineers are notoriously fussy if they feel the pilot is hard on their baby. I don't ask to see Ken's log book. I'm not that cheeky, and besides, remember Jacques back in Chibougamau in '62? Jacques had umpteen hours in his log book and we had to build a fire so he could find us!

I tell Ray he is going to train a new guy tomorrow, and Ray rolls his eyes. I don't care - Ray will have to suck it up. I know he will take care of the kid and Ray has a lot of good bush sense. (Not so much in town however, an unfortunate yet interesting trait of the bushwhacker genre.)

The next morning the kid shows up sporting a new fluid filled compass and a decent looking staking axe – so far so good. The chopper takes Art and his crew in first. The next two trips will drop off Ray, Tim and the kid. Ray and the kid go in first and Tim takes the second trip.

(Now let's take a minute to recall the first rule of chopper work. Two men in followed by one man; one man out followed by two men. Never, never, leave one man in the bush alone, right? Ok, read on.)

With the men safely in the bush I return to my room to do some paper work and chew on my favorite doggy treat, the ever-present financial bone. I don't know it yet, but soon events will occur that will remain indelible in my memory forever – events which epitomize the true spirit of the North Country. Do your job and do it well and if you have to go that extra mile don't brag on it.

After lunch the weather turns on us. A fast moving front rolls in from the west and by 2:30 pm it is snowing so hard I can't see across the street. I drive over to Central Pat and find Ed starting supper at the Legion and I tell him to hold on for a bit, the boys might be spending the night in the bush. Ken comes down from the hotel and he is apologetic and a little frustrated at his inability to fly. I tell Ken not to fret about it. These boys know the ropes and they will be ok. Art and his men will probably walk out anyway, and they do. Ed has been patrolling the road since 5 pm, and at 7 pm he picks up Art and crew and they return to the Legion, tired, wet, and hungry as bears, but happy to be warm and cozy for the night.

At 8 pm I am having a cup of coffee with Art when the dam chopper starts to warm up. I whip outside to find that Ken and the engineer have pulled the covers off the G4A and Ken is going to find Ray. The front has passed through and the full moon on the fresh snow is so bright you can almost read a newspaper outside. No matter how hard I try to convince Ken that Ray and crew will survive quite well around a campfire, he is adamant. His job is to bring the men out and bring them out he will. The engineer tells me Ken is as stubborn as a mule and I might as well let him go, so we watch the chopper fly off into the darkness.

Fifteen or twenty minutes later Ken returns with two passengers, Ray and Tim, and my heart drops into my boots. Where is the kid? Ray climbs out and he is almost crying. The kid headed north three hours ago and no amount of argument short of physical violence could dissuade him. The kid told Ray that it was only two miles to Pickle Crow and he knew the country. There was no way he was going to spend the night in the bush, and that was that. Ray had to let him go.

Ken is going to find the kid and Ray goes with him – Ken needs a spotter. So off they go into the darkness again.

Now, the kid might have had half a chance if he was on snowshoes, but it's always tough to make that transition in the early winter. Ray's claim block is on relatively high ground and so far snowshoes are not quite necessary. This snowfall will tip the scales and tomorrow everyone will strap on the Chestnuts.

The swamp leading up to the old mine site of Pickle Crow is a big one with no open water to speak of, but lots of hummocks with deep pools of unfrozen water in between. They find the kid half way to the Pickle Crow road, wet to the ass, half frozen, and almost comatose – just sitting on a hummock, totally whipped and waiting quietly for certain death from exposure.

We load the kid into the pickup and I haul him home, tell Mom to throw him in a hot shower, and explain the situation to Dad. I find out the guy is only 15 years old! Holy Cow! What were they thinking?

The next day Dad brings a crestfallen and contrite boy around to apologize. I offer to pay him a day's wages, but Dad refuses. He tells me no amount of compensation can ever pay for the lesson his son has learned. What's the kid going to do now? "He is going back to school." says his dad, "Maybe now he can learn to pay attention."

I've always figured Ken saved more than one life that day. Good on you Ken Venturi, wherever you are.

Either Lady Fortune decides to reward us, or more likely because the puzzle pieces are starting to fit in, from that day on everything smoothes out. We are almost done by now and Abe is keeping me even with the bank.

The last of my stakers pull out, and we wave goodbye to Venturi and his engineer. Abe has now paid me for everything except the last 100 claims. I have settled up with all my stakers, and I clear up a few minor local bills. Ed and I clean the Legion, return the borrowed beds to Inco Bert, and head home for Christmas on December 23rd. On the way home we swing by Sioux Lookout to record the last of the claims.

What a relief it is to be home! Lots of rugrat cuddles, and a decorated tree in the living room is loaded with presents. I can hardly wait for Christmas morning to see what I have given the family – Santa Claus has had to do my shopping.

On December 27th I send the last of the claim transfers to Montreal, along with my final invoice. I contact Abe after January 1st, and he tells me he is flying to Dryden on January 10th to give me my final payment. I meet him in the bar at his hotel, and receive his thanks and a cheque for $10,000. Both are much appreciated, and just like magic, my left eye stops twitching, and my tummy pains are gone. The Umex rush is over!

I don't even rush to the bank – I have no doubt the cheque will clear. I want to take it home and lay it on the kitchen table so the wife and I can admire it for a couple of days.

Sidebar: Although I was never able to follow up on Abe's success at the time, I believe he dealt some of his claims. Twenty years later we had a C-store/gas bar/restaurant in the Rainy River Valley, and one of our customers was an underground miner, working two weeks in/two weeks out at a gold mine just south of Pickle Lake. From his description of the mine location, I deduced that it was located on a 99-claim block we had staked east of Hwy 599. The mine had been in production for more than ten years, and was now on its last legs, but with the combination of cash, stock and a 3% net smelter return, Abe would have been well compensated for his staking expenses.

Table Scrap

Bill and Dave had staked roughly 220 claims, and most of the remaining 708 were staked on my licence. I had to do it that way for more than one reason. Contract stakers using their own licences must sign a transfer form assigning the claim ownership to the contractee (Abe.) I didn't know most of my Hwy 599 stakers that well, and had any of them gotten wind of my shaky finances, they may have refused to transfer claims before being paid.

There was also the problem of the recording office in Sioux Lookout being at quite a distance – over an hour by air, and at least six hours by road. I just plain couldn't trot in to buy tags for every staker, but a couple of sked flights kept my own supply of tags topped up.

Now here's the sticky part: When staking, the date and start time must be written on the #1 post: Nov. 15th, 8:30 am, for example. Each #1 post indicates that particular claim's start time, and you can start at 7 am and run as late as you wish, but in the short winter days 5 pm is pretty much the max. Usually you space the times 1 to 1 ½ hours apart, depending on the heaviness of the bush being staked. If the claim is beneath a lake, then witness posts on shore indicate ground claimed, and times can be compressed. Three men using three licences can space their times quite comfortably, but at least three crews are using my licence – thus my start times have to be closer together. I am spacing my start times at intervals as short as 15 minutes, while making sure that the dates don't put me in two places at the same time.

Had anyone disputed my claims I would have had to prove to the claims inspector that I could perform as advertised. It is perfectly legal to hire men to blaze the lines and make the posts but I have to walk (run) the one-mile circumference of a 40 acre claim and affix the tags myself. It would be equivalent to a 4-minute mile for each claim – and Roger Bannister himself couldn't have done it. I was playing with fire every day and it was just dumb luck that I wasn't caught.

Table Scrap: The Hare Crews and the Tortoise Crew

This is told by someone who generally tells more lies than me – thus, while it may not be true, it is entirely possible.

It was the 1960's in the Goganda/Shining Tree area west of Temagami. A hot four-claim block was coming open at 7am and three groups wanted the ground. Each crew consisted of four men, and each man had his own staking licence and tags for one claim.

The groups had arrived on site the day before. Group one had a chopper, group two a fixed wing on floats and group three a beater pick-up.

Now here's the deal, and it's kind of neat.

This will not be your typical every-man-for-himself rush. The staking cannot begin before 7 am sharp, and with three men at every Number One post, no one can get a head start. The rush is to get to the recording office – first to record in Sudbury will get the prize.

So they all work together, and then rush to their respective transportation choices. The pickup boys see bad weather coming – they may yet have a shot.

The fixed-wing won't take off. His base in Sudbury tells him they are socked in and there are few lakes between Shining Tree and Sudbury for an emergency stop.

The chopper leaves, hits the front and lands in a swamp to wait it out.

The pick-up boys rattle-trap it out on the turkey track to Temagami, thence 50 miles south to North bay and 90 miles west to Sudbury, hoping the pistons stay in the holes and stopping often to top up engine oil. They arrive at the recording office, record their claims and meet the chopper crew coming in as they are going out – Chopper Guys were 15 minutes late.

The fixed wing had a longer nap and arrived a full day late and also four claims short.

Table Scrap: How to Reorganize a Union Organizer (as told to me, and later corroborated by one of the culprits)

It was Pickle Lake in the mid-60s, Ted Ankney was teaching school at Central Patricia, and Gordon Franklin was working as a millwright in one of the mines. (They later teamed up as A&F Logging – see Chapter 8.)

A union rep blew into town, checked in at the Pickle Lake Hotel at the Landing, set up a "throne" in the bar and proceeded to piss everybody off. He was loud, obnoxious, and openly disdainful of the local non-unionized hicks – but he bought the rounds.

The two mines treated men fairly. After all, being this far off the beaten path, there was not much of a labour pool to work with. If you had a good work force, you wanted to keep them happy.

Ted and Gordon were having a beer one evening and listening to the union spiel. The guy had been there a couple of days and he was rapidly wearing out his welcome. The union rep staggered off to bed just before closing time, which meant 11 pm in those days.

T and G decided to yank his chain a little, and of course, we all know that the most sensible plans are drawn up over a few beers.

The Pickle Lake Hotel was a fairly large 2-storey building of a square design. As you walked in the front door you were faced with a broad stairway leading straight up to the second floor. The rooms were situated around a central open area and the union rep's room was to the left of the open staircase. The beds were the old style with iron head and footboards, and nobody ever locked their doors.

Gordon, for some reason, had a 200-foot coil of heavy rope in his van, probably used when they re-strung the hoisting cables at the mine.

They snaked the rope in the front door, up the stairs, and tied it to the foot of the union man's bed, with the other end tied to Gordon's van.

"Give him a little jerk," says Ted, thinking this would be good for a laugh the next day.

Gordon jumped in his van, put pedal to the metal and drove home.

When the bed hit the door, the footboard came off, and so did the door frame. The footboard did a couple of circles, scooted down the stairs, took out the front door on the way by, then headed down the road behind Gordon's van in a cloud of dust - followed by Ted.

As other guests rushed to their doors, they saw Union Guy at the head of the stairs, mattress and all. As he climbed out of bed in bewilderment, he hollered "Earthquake! Earthquake!"

Within half an hour the union rep was also raising dust as he headed down 599 – never to return.

Gordon and Ted hunkered down for a few days, the owner of the hotel never did find out the whole story, and the mines remained unionless.

Table Scrap: Payday

The Central Pat Hotel was managed and in the process of being bought by Bob Parker. I had worked with Bob in '66 at Camp 20, south of Smooth Rock Falls, near Cochrane in North-Eastern Ontario.

Bob was a good guy with a sharp sense of humour. At that time (and perhaps still?) it was mandated by the Liquor Control Board of Ontario (LCBO) that a sign be posted stating the seating capacity of any bar. The sign must read, "Any more than __ persons in these licensed premises is dangerous and unlawful."

Bob added his own sign in a prominent place above the shuffleboard table: "Any more than 56 persons in this licensed establishment is PAYDAY!

Chapter VII   Savant Lake: Winter '71

As I drive home that day with Abe's ten grand in my jeans I can feel that money starting to burn a hole in my pocket. I'm going to put some of it to work, and plans have been in the works for a few weeks already. I am on the threshold of becoming a major player in the North-Western Ontario mining exploration community.

Before Christmas, and before he left Pickle Lake, Bill and I had had some primary discussions regarding our mutual futures. Bill and Dave were planning to split and Bill thought he and I should hook up. He was impressed with my ability to handle my rag-tag bunch of stakers, and my geophysical skills. I was impressed with Bill's work ethic, reputation, and wide range of contacts. We would make a good team – at least in theory.

We have been keeping the phone lines hot since Christmas. Bill has been negotiating with Don and Peter and they have come to an agreement. Bill and I will form a company and we will contract to do all Noranda's staking, geophysics, and most of the line cutting. Some of the larger line cutting jobs will be done by the Val d'Or boys for now and we hope to take over the bigger jobs as we progress. I have been talking to Ray and Rod and the table is set. Ray and Rod will finish the job they are on right now and will transfer to our pay roll before the end of January. Peter will continue to run one crew out of Thunder Bay for now and we will absorb them in the future.

I use 5000 dollars of my UMEX rush money to open a business account at a bank in Dryden and I head to Thunder Bay to sign the contract. Bill is to match the five grand in his own business account in Thunder Bay. We will each handle our own end of our two departments, me with the Noranda contract, and Bill with staking contracts and prospecting endeavours. At the end of the year we will do the books and split the profits. Yeah, right.

I sign the contract, but I have a few qualms in doing so. Bill and I have a private discussion before the signatures are put to paper and he thinks I am being a worrywart. He is willing to overlook some of the grey areas. Bill has dealt with Don for years, and Don has always been fair regarding exceptional (unforeseen) expenses. What Bill fails to take into account is that Peter has been instrumental in the formation of the contract, and he doesn't know Peter. I do.

Basically, the contract calls for X dollars per mile or per duty performed, with a sliding scale kicking in on line miles per job. In other words, the larger the individual job, the less dollars per mile. This makes sense for both parties; moving camp requires a day or two of unproductivity and the larger the individual job, the less moves. I want a mobilization/demobilization price in there, but Peter resists the concept. "Trust them," is Bill's response, but I do not trust Peter, and I will be proven right.

122

CONTRACT

BETWEEN   NORANDA EXPLORATION COMPANY LIMITED,
                                          (No Personal Liability)
                                 C/O Peter Gordon Cooper,
                                 253 Lincoln St.,
                                 Thunder Bay, Ontario.

HEREAFTER REFERRED TO AS THE COMPANY AND

                                 Gulchs' Vulters Company Limited,
                                 C/O Bob Durnin,
                                 Vermillion Bay, Ontario.

HEREAFTER REFERRED TO AS THE CONTRACTOR.

    The Contractor shall do Electromagnetic and Magnetic Surveys, where and when required by the Company in the province of Ontario and other areas as may be necessary from time to time, and in accordance with the provisions set out hereunder.

1)   The Company shall supply the necessary geophysical instruments for the Electromagnetic and Magnetic surveys.

2)   The Company shall pay for the necessary repairs for these instruments except when they are willfully abused by the contractor and or his personnel.

3)   The Company agrees to pay for two flights on all fly jobs from an airbase nearest to the property in question. Properties with 50 miles or more will require at least one additional flight but the number of flights must be agreed upon before the property is surveyed.

4)   The Company agrees to pay for the surveying of a property with a cut grid of:

    (a) 5 miles or less $40. per line mile of Magnetic Survey
                               $60. "   "   "   " Electromagnetic Survey

    (b) 10 miles or less $40. "   "   "   " Magnetic Survey
                                50. "   "   "   " Electromagnetic Survey

    (c) 15 miles or less $40. "   "   "   " Magnetic Survey
                               $40. "   "   "   " Electromagnetic Survey

    (d) 15 miles or more $40. "   "   "   " Magnetic Survey
                               $35. "   "   "   " Electromagnetic Survey

5)   The Contractor shall take readings every 100 feet on high frequency. In the case of anomalous zones the readings shall be taken every 50 feet with dual frequencies.

6)   The coil spread shall be 200 feet except over lakes and other areas where deemed necessary the coil spread shall be 300 feet.

7) The Contractor shall supply field notes and field maps for the Electromagnetic and Magnetic Surveys indicating all readings plus topographic notes, claim boundaries, and claim posts where possible.

8) The Contractor shall be responsible for all necessary medical, hospital, unemployment insurance, workmen's compensation, Canada Pension Plan etc., for all his employees.

DATED AT THUNDER BAY, ONTARIO THIS _____ DAY OF _____ 1971

WITNESS _____

BOB DURNIN _____

_____
PETER G. COOPER, AGENT FOR
NORANDA EXPLORATION COMPANY LIMITED.

This winter we will be following up on airborne responses in the Sturgeon Lake/Savant Lake area, so I head home, give the Impala back to my wife, load up the Pontiac and head for Savant Lake to prepare for the arrival of Ray and Rod, who I expect in a day or two. I check in at the Savant Hotel, have a bite to eat, enjoy a beer with Bob Dunham (the owner/bar tender) and hit the sack. It's a cold, cold night, so before I go to bed I plug in the Pontiac's block heater and interior car warmer.

I awaken the next morning in a dark and rather cool room. I grope my way down to the dining room and Mrs. Bob tells me breakfast will be late. They have trouble with their power plant and Bob is working on it.

Sidebar: Ontario Hydro had no line into Savant. The new mines south of Sturgeon Lake would soon facilitate pole construction from Ignace, but the line will reach Savant two years in the future. In 1971 the general store, hotel, and CNR station/company houses, had individual power plants. Private homes had a choice. Nick, a local entrepreneur and all-round good guy had a huge old babbitt-bearing single cylinder power plant resurrected from a Saskatchewan grain elevator, and he supplied power to a number of homes. Others had to do with coal oil lamps and wood or propane stoves and heaters.

I cool my heels in my room for a couple of hours. The lights come on, and breakfast is served. After eating I head outside to start the Pontiac and when I open the car door I am greeted with the bitter acrid smell of burnt synthetic carpet, and my interior car warmer is a misshapen blob of melted plastic. Uh oh, guess who killed the light plant?

That afternoon in the beer parlour Bob tells me that whatever the problem was, it must have been quite a sight in his generator building. He says it looks like a lightning bolt had hit and his main breaker panel was blown off the wall. He has made temporary repairs and an electrician will have to come in and it's going to cost him X amount of dollars, and he knows that the problem isn't in the hotel wiring, and on and on.

I keep my lip zipped. I know my car warmer is the culprit, but the deed is done and I don't want to get off on the wrong foot here. There is still three months of winter left – plenty of opportunity for me to make the odd donation at Bob's bar.

Peter arrives the next day with Ray and Rod. They bring their personal stuff and the geophysical gear with them; I have the 12x14 tent and associated camp gear in the Pontiac.

After lunch we retire to the bar for what you might call a transfer of power ceremony and we raise a glass to mark the occasion; from this point on Ray and Rod will be working for me. Good luck to all of us.

A charge account has been set up at the general store next door and I send the boys over to put a grub order together. Peter and I go to my room and he brings out the maps.

As soon as I see the maps I know I am in trouble. Peter has us working on blocks of four claims! I immediately point out that I can't make a profit on four claims, and Peter pretty much tells me to suck it up, I signed the contract. Well he's right about that – sign it I did, and there is no point in complaining about it now. My only hope is that we can make it through the winter in one piece. Next summer the terms can be adjusted to a more workable deal. Maybe, just maybe.

I have a multitude of problems on my hands here, and I'll try to explain them using my past experience as comparison.

Anomaly chasing can be an iffy proposition. You have to have some confidence in the airborne contractor in regards to the accuracy of their plotting, but errors can occur. Inco's search square system, while only covering an area of 1600 feet square, still allowed for further expansion of the search block if results warranted an enlargement of the work area, and Inco never worked out of a snow bank camp in the winter time. Men were flown in daily from a base camp.

Mining Corp and Noranda seldom worked on blocks smaller than 16 claims in the summer, and in winter the jobs were more likely to be 20 claims or larger, justifying a more elaborate and comfortable camp, and besides that, the claims were already staked and the lines were cut and chained before the geophysical crew came in.

Now, claim staking is not rocket surgery. Even with minimal experience it's not hard to stake a four-claim block. One man can handle this in one day without killing himself, two men can do the job in 4 hours. After staking, four miles of line has to be cut and chained. (It's a basic rule of thumb – 1 line mile per claim). You follow up with four miles of geophysical work on each instrument – four miles EM, four miles MAG – then you move camp and the whole process begins again.

There is another fly in the ointment. Line cutters are line cutters, and geophysical guys are geophysical guys, and seldom the twain shall meet. Geophysical men can bang off four line miles of survey a day, but few can cut more than 1 mile a day. We should never have included line cutting in our contract. That opened the door for Peter to penny-pinch us to an early death.

I'll use the first job as a case in point.

The next morning the three of us fly in to the first 4-claim block. A 180 from Sioux Lookout comes in to the little lake just south of Savant and hops us the 15 miles in two trips.

It takes us seven days to do the staking, line cutting and geophysics on four claims, 4½ days to do the actual work' and 2½ days to move in and out. (We sat in a snow bank for a day waiting for the weather to clear before the plane came back.)

I can see that the contract as written leaves little room for profit. Expenses are virtually equal to income and none of us are making much more than day wages. I have put Rod and Ray on a base wage structure with a production bonus, but on a 4-claim block they will never reach the bonus level. They could make more money had they stayed with Noranda, and added to the unworkability of the setup is that proper maps are needed to send to Thunder Bay, and it's impossible to generate a decent map in a snow bank tent. We will have to go to plan B.

I send the two guys in to the next grid with the assurance that I will hire a third man to help them, and I do so. Gene joins the crew a week later. I will remain in town to do the map work and will try to generate more business of the profit-making variety. The stage is set for what will prove to a very interesting winter, and in years to come I will often think of the Chinese curse: "May it be your bad fortune to live in interesting times."

I rent a small house from a local semi-bandit and this will be my office/bunk house. It contains a minimal amount of furniture, but there are a couple of beds, a sofa and a small kitchen with a table suitable for map work. Now I won't have to pay for hotel rooms when the men come in and out, and expenses will be reduced. The house is a little too close to the bar, but what can you do? It's a very small town.

Right off the bat I go to Pickle Lake to stake 60 claims on spec. I have been keeping an eye on things up there, and a good sized area right in the middle of the action is still open. Maybe I can make a buck or two, so I hire Rosaire (part of Art's crew from last fall,) and we head for Pickle Lake for a few days. It doesn't take long to stake a big block right on Pickle Lake itself. Most of the claims are under water, so I buy a number of 4x4 timbers, cut them to claim post length and we plant them along the shoreline (some of the posts are pretty much in people's front yards.) About ¼ of the block extends in to the bush just north of the lake, but the area is virtually treeless. We give 'er tarpaper and return to Savant four days later. I will wait for offers to roll in – I am already in clover (in my dreams.)

I put out a few feelers, and it doesn't take long to find out why the Pickle Lake ground was still open. The claims are worthless. The UMEX rush is over. It's just like my attempt to burn Fecteau's Beaver 10 years ago. The flames had flared and quickly died, and I am land rich and cash poor.

Eight months later I receive an offer from UMEX, of all people! They are building a road into the mine and one of my claims has some good gravel that they need. My claim does not entitle me to the gravel rights, but they want to avoid hassles. I gladly sell them one claim out-right for $900. I almost break even, and to me that is a victory.

The winter goes on, and despite my problems with Peter I am enjoying life in Savant Lake. It's a great little town with all the small-town quirks, and I just love the place. This is frontier country for sure, and law enforcement is virtually non-existent. There is an OPP detachment at Pickle Lake, 100 miles north, another at Ignace, 100 miles south, but for some unknown reason Savant Lake is policed from Armstrong, and Armstrong is two hundred miles east on the CNR north line. You can drive to Armstrong, but you'd better pack a lunch before you leave; you have to go to Thunder Bay and head north 230 miles up the Spruce River road. The CNR passenger service is notoriously unreliable, and a call to the Armstrong detachment might result in an officer stepping off the train a day later after the miscreants are long gone. You will see a black and white occasionally on Hwy 599 as they motor in and out of Pickle, but they seldom take the ½ mile detour in to Savant, so Savant is essentially self-policing. It's no big deal. These folks are a pretty much laid back group, and if problems arise they are taken care of locally, no lawyers or remands or all that crap. Occasionally someone gets thumped, but it is seldom necessary. I will spend the better part of four months In Savant and I'll never see much more than a loud argument.

It's really nice that I have an old friend in town. In the late 40's our families lived on adjoining farms in Manitoba. Lloyd is making a career of it in the CNR Signals Dept. He and his wife are living in a company house and I get the odd free meal. Lloyd and I tip the odd glass in the bar and reminisce – we remain friends to this day.

Table Scrap

The westbound passenger train is scheduled to hit Savant at 2 pm. One day it arrives bang on at 2 pm exactly. That evening I mention to Lloyd that the service is improving and he laughs like crazy! "That was yesterday's train," he says, "It was exactly 24 hours late."

127

Table Scrap

Every town has to have a feud, right? Well, it's not the Hatfields and the Coys, but get this: The station agent, for some reason is mad at Nick, the power plant guy. The problem is, the station agent needs 100lb propane cylinders for his kitchen stove, and Nick is the guy who fills the cylinders. So if station agent guy wants propane he tells Emerson Ennis at the general store, who calls Nick. Nick brings the 100lb tank to Emerson, Station Agent Guy walks across the street, exchanges his empty tank for the full one and pays Emerson. Nick now comes back to the store to pick up the empty, and Emerson pays Nick. The transaction is complete, and by using Emerson's store as a demilitarized zone, faces are saved on all sides – wonderful! I feel like giving Savant Lake a big bear hug.

Table Scrap

How about this one? One afternoon I am taking a beer break and trading stories with Bob, the bar owner. It's a quiet afternoon and the only other customer goes into the men's rest room facility. He returns to his table and there is a muffled <u>whump</u> and white dust and smoke pours out from the space between the rest room door and the floor. It startles the dickens out of me, but Bob never misses a beat, just finishes his turn at story telling. I ask him what happened, and Bob opines that the ceiling must have fallen. Over the years the rest room has been spruced up from time to time, and rather than remove the old drywall, new sheets have been added over the old. Bob says he has been expecting it to give way anytime. I have to have a looksee, and when I open the door I find four inches of broken sheetrock covering every square foot of the room, including the urinal and throne. Had it happened two minutes earlier it would have surely driven the other guy to sobriety. Is the bar business interrupted? Heck no – we just use the ladies can.

Table Scrap

Another day and another beer break: Today there are no other customers, and things are pretty quiet. Bob and I are having our usual BS session, but as we chat I'm picking up some strange vibrations. It sounds like someone is retching and it seems to be coming from beneath my feet. Bob seems unperturbed, as usual, but I am getting very curious, so I ask him what is going on down there. He tells me it is just _____ and _____ who are cleaning his septic tank. Bob has a tile bed across the street and his primary settling tanks are in the basement. When the excrement has broken down a transfer pump moves the liquid sewage to the tile bed for evaporation. Once a year the covers are removed from the tanks, and in exchange for free beer, _____ and _____ shovel sludge into 5-gallon pails, which they carry out the rear door of the basement and dump into a honey wagon for a trip to the landfill. Bob says it's fairly expensive. After an afternoon of puking their guts out _____ and _____ need many barley sandwiches to reline their stomachs.

My expenses are starting to exceed my income at an alarming rate. For one thing, it seems that I am doing a lot of free work for Peter. Roland Sevigney (Black Beaver) has brought in a line cutting crew from Val d'Or to put a large grid in at the north end of Savant Lake (the lake itself not the town,) and although I help them move in it turns out that my men will not be doing the survey – Peter will be using his own crew.

It has become evident to me that Peter is building me a coffin. At first I thought that his lack of experience was behind the unfairness of my contract, but now I'm beginning to see the light. Peter is jealous of my success since leaving Noranda, and resentful about the Noranda contract. He is an empire builder and there is no room for an independent block in his pyramid. The moment of truth, so to speak, arrives in early February.

Sam is coming in with his drill, and his first hole is 10 miles up 599 and 2 ½ miles east. A Noranda field geologist has spotted the hole and has marked the way in with surveyor's tape tied to tree branches. A road has to be cut through the bush to the drill site, and Peter wants to know if I want to cut it and what will it cost him. I tell him I have to do some pencil whipping and I will call him back.

I do some figuring. I know a thing or two about a drill road, and it's not exactly an easy task to make a decent road through virgin timber. The road must be wide enough for Sam's muskeg tractor, and while it doesn't need to be arrow straight, sharp curves are unacceptable. Trees and clumps of tag alders and willows must be cut at ground level, and the snow is deep this winter. In some spots downed trees will be cut into smaller logs to fill in the rough spots to make the road a bit more level. It's all bull work; chain saw, shovel, axe and sweat. My men will be moving camp in two days and I will hold them over in my house in Savant. If we keep the pedal to the metal we can finish the road in two or three days.

I tell Peter we will do the job for $800, and he lays the hammer down. I'd damn well better have the road done in less than a week. Sam will be coming in then, and Peter doesn't want to pay stand by time, and don't let him down, and on and on and on. Well, la de dah!

That evening I have a beer with Ted and Gordon, the local logging contractors, (remember the union guy?) and we have something in common. Over the past couple of years they have been friends with Carl, my old North Carolina/well drilling buddy, and I have been a beneficiary of that friendship, sort of an honorary member of the club, if you get my drift. Anyway, they ask me what's new, and I tell them about the road. Two rounds later Plan B takes shape.

Tomorrow morning they are floating a D8 dozer to their timber limits north of Savant and they are going right past my road job. They will offload the D8, push my road in, reload the cat and be on their way. I will pay them $100 per hour and Peter will have his road. Deal. We raise a glass.

Bright and early the next morning I am watching the operator unload the dozer. I tell him to follow the ribbons and I tag along, taking care not to get in the way. At the end of the road I make sure he leaves the ribboned picket marking the hole undisturbed, and I follow him out.

One pass in and one pass out, with a little backing and filling to smooth out some rough spots, and 5 hours later the dozer is on its way. Sam has a class 'B' highway to the drill site and I am happy as a pig in poo.

I hustle my buns back to Savant, pay Ted 500 bucks and call Peter with the good news. Is he happy? No he is not. Peter loses his cool and I can only listen in disbelief as he hollers at me. He tries to do me favour, he says and I pull something like this?

What the hell is wrong with this guy? He is yelling at me about me how I'm ripping him off, and how he could have hired the cat himself, etc. etc. I point out a few facts to Peter. Even if he had been able to make the necessary arrangements with Ted and Gordon he was not on the spot at the time, and even if they were willing to pull the D8 off the job at their timber block, Peter would have had to pay for floating time, and the total costs would certainly have exceeded 800 bucks.

No amount of arguing will change Peters mind. He will pay 500 bucks, and that's that. Peter's word is worth so much toilet paper, and once again I am working for nothing. The final nail is poised above my coffin.

Peter is not the only carpenter I have to deal with. I am also having trouble with my partner. I look back on those days and I can only blame my inability to handle Bill on hero worship. For years I have been hearing stories about the exploits of Bill and Dave, and I had thought that I was fortunate that Bill wanted to tie up with me. Now I am starting to think that unfortunate may be the operative word.

I am not seeing any income from Bill's end, only expenses. Is he generating any cash flow? If he is, I sure don't know about it. Did he ever come up with his 5000 bucks? Again, I don't know about that either. All I know is that I'm writing checks and Bill is cashing them. Bill has an Aeronca Champ, and he has been extolling the benefits of having our own plane to facilitate his prospector endeavours, and how the Champ can save us charter costs. Well, so far it has not saved me any money – quite the opposite. Bill wants to replace the Champ's engine with a 90 horse power plant to give us STOL capability. I write a cheque for the new motor – 2000 bucks! We are saving <u>Bill</u> some money aren't we? Bill wants to hire a young geologist to prospect with him. I pay the guy's wages. Bill wants me to hire his nephew. I hire him and put him with Ray and Rod on a road job we are doing south of Savant. The kid is a smart ass with an over developed sense of entitlement, and is a clubhouse irritant – a waste of money.

Bill flies to Savant in his newly powered Aeronca and picks me up. We are going to check out a showing on an island in a large lake halfway to Sioux Lookout. It's early April and warm weather has been eating on the lake ice, but it was pretty cold last night and Bill figures the melt water on the ice has frozen enough to support the Champ. Bill is wrong. The ice looks ok from the air, but when we land the Champ breaks through the two inches of fresh ice into 12 or 14 inches of water. The lake ice is not yet candled and the ice beneath is thick enough, but the surface ice shreds the fabric on the Champ's under belly and does some damage to the tail section.

We step out into cold water over our boot tops and survey the damage. Bill figures he can take off without me and make it back to Thunder Bay, so we break ice to give him enough runway to lift the aircraft out of the water. He says he will call Sioux Lookout to come and pull me out and I watch him take off, leaving me alone, wet to the knees and with nothing but an axe and a useless pair of snowshoes. It's pretty darn quiet after Bill leaves, and I ponder my fate: I have broken my own rule – never leave one man alone in the bush.

I keep busy for the next four hours. I figure if a plane does show up, he needs half a chance, so I break almost half a mile of ice, making sure to hack the jagged pieces into small bits so as not to harm the aluminium skin on my rescuer's aircraft. My legs feel like two frozen stumps, but I do some rough time estimations. If no one shows up before 4 pm I still have enough daylight left to build a fire and a small shelter on the island.

Slate Falls Airways, bless their pea-pickin' hearts, comes through in the pinch. At 3 pm a 185 shows up, shoots up the lake, spots me and my crushed ice landing strip and pulls me out of there. I feel like giving that pilot a big wet sloppy kiss, but he acts as if it's just another day in the life of a bush pilot. He drops me at Savant. I get to change out of my wet clothes and I have a good story to tell in the bar that night.

That is pretty much the end for Bill and me. I pay for the 185 charter, but refuse to re-fabric the Champ. Enough is enough. Things wind up with more than a whimper rather than a bang, though. I slowly distance myself from Bill and Noranda. Snow conditions render further work on the road job impossible, and I lay everyone off. I never do get paid for the portion of work on that last job, and I never speak to Peter again. Rod returns to Noranda where he stays for many years. Ray goes on to drive truck and will meet an untimely end. I hit the independent bush trail again.

Table Scrap

Ray always wanted to be a trucker like his dad, and had saved a few bucks. When I laid him off he put a down payment on a rig and hit the wood haul. I ran in to Ray a couple of times in the seventies, but he was having a tough go of it. Bad luck was the norm with Ray it seemed. I heard he had gone on to drive for wages on the Prairies. In 1982 I was working at a pipeline terminal in Winnipeg and one day I saw an item in the newspaper about an inquiry into a fatal industrial accident. A trucker, delivering anhydrous ammonia had been working on a hose fitting underneath his trailer. The fitting had come apart, dumping anhydrous on his face and he had died instantly. The trucker was Ray.

Table Scrap: Karate Kid (As told to me by Carl Haglund)

Stan Werbiski was a successful Pickle Lake/Savant Lake/Sioux Lookout businessman. A very nice person, his only fault – if you could call it that – was his susceptibility to having his chain yanked. He would believe anything you told him.

Carl was holding down his office chair in the Savant Lake Hotel bar along with Ted and Gordon. Stan usually stopped in on his way home to Sioux Lookout every Friday. (At that time you had to drive down Highway 599 to Ignace, then over to Dinorwic, and back up to Sioux Lookout.) He would also always stop in for a quick beer – and the boys had set him up.

The conversation was steered around to the subject of Karate. Ted mentioned to Carl that he had heard that Carl had his black belt. Carl said he didn't like to brag about it, but he had been known to break four or five one-inch boards with the edge of his hand.

"Ha ha ha," said Stan, "I know you're lying this time."

Now, Carl was 5'10" and probably weighed about 160 pounds if he got caught in a monsoon.

"It's true," said Carl, "It's all breath control and channelling your energy properly. Why, I could bust this bar table in half if I wanted to."

(These were old-time bar tables, made of wood, with 2 1/2 inch cross-pieces between the legs and with 1" boards on the top.)

"You're BS-ing me again," said Stan, "Let me see you do it."

"I already broke one," said Carl, "Bob (the owner) would be really mad if I did it again."

"Go ahead," said Ted, "I'll pay for it."

So Carl got up, flexed his muscles (??) and lined up for the blow. Bob was waiting around the corner of the divider to the Ladies and Escorts side. As Carl lined up, Bob came around the corner and yelled, "Carl, you son-of-a-bitch, if you bust one more table in here I'll cut you off for life!"

I don't know if they ever did tell Stan the truth.

Table Scrap: Helping out

Carl Huston, a geologist from Red Lake was drilling about ten miles south of town, not far off Highway 599. I had been introduced to him in Red Lake during the Uchi Lake rush in '69 but did not know him well. He was a decent sort.

Now, you must realize that in the mining exploration game, everyone has their ups and downs. I think Carl was going through a lean time. He had rented an old 8x40 house trailer, probably a hunting shack, just off Highway 599, and was drilling with a Winkie drill on his own. That's a tough proposition at any time, and almost impossible in the winter. You have to keep the water warm with a coil heater, and once the drill is in the ground you can't stop until the hole is finished. Carl had been running the drill alone 24/7 – how he did it, I don't know.

He popped into the bar one day while I was there, so we shared a beer. "What's up?" I says.

He told me he was waiting for the passenger train, as he had hired a driller from Sudbury and expected him on the train, which was late as usual. So Carl checked in at the station, came back and said he had to go back to camp and check the water pump. He asked if I would watch for his driller if he came in.

"Sure," I said. Carl did not look at all well – stressed right out.

The train showed up, and the driller staggered off, and he was hammered! I took him to the bar to wait for Carl, who showed up shortly after. The driller had to have another drink, then Carl, pretty cranky by now, loaded the driller into his old Ford Econoline van and headed for camp.

One more beer for me, and back in came Carl and his driller. "Oh-oh," I said to myself.

"I rolled the van about two miles south," said Carl "Is there a wrecker (tow truck) in town?"

"Nearest one is in Ignace," said I (90 miles south.)

"How am I going to get a wrecker at this late hour?" said Carl, "And I have to get back to the drill before it freezes!"

"I'll run you back to the trailer and we'll deal with it in the morning," I told him, and we jumped in my car and headed down 599. By this time I was feeling really concerned about Carl. Dejected and downcast, he looked like he didn't have a friend in the world. We got to the van and I said, "This doesn't look so bad. I don't think we need a wrecker at all."

The winter of '71 was an old-fashioned one, cold as hell and lots of snow. Carl – in a bad mood and driving too fast, had lost it on a curve, and laid the van on its side in the ditch, in the soft snowbank.

Now, I knew that Ted and Gordon were moving pulpwood, using hired tandems, to the rail siding on the CNR Thunder Bay cut-off 40 miles south, and I knew they were running around the clock.

So we turned around, and at 10 o'clock on a cold winter's night I was knocking on Gordon's door. I explained the situation to him, and without hesitation, Gordon took over. He put the two of them in his pick-up, grabbed a chain, stopped the next tandem and yanked the van back on its feet. Total damage - one broken rear-view mirror. The engine didn't even lose any oil and Carl and his driller were off to camp. The next evening I dropped into Gordon's and asked him how much Carl owed him for the favour.

"Bob," said Gordon, "In this part of the country favours are calculated in multiples of 2-4's." I figure this is a 2x2-4 favour – one for the trucker and one for me."

At that time 24 beer cost a little less than six dollars, so I delivered it the next day. A few days later Carl ran into me in Savant.

"How much do I owe that fellow?" said Carl.

"Taken care of," said I.

Ten years later I was living in Winnipeg and took a two-week working vacation to do a short geophysical job in the Atikokan area for an old geologist friend. I got a cabin at a tourist camp on the Seine River and met a young fellow there who was also doing a job on an adjoining claim block. It turned out that he was the son of Carl. He called his dad and mentioned that I was in the area.

"You take good care of that man," said Carl. "He saved my life ten years ago."

Table Scrap: Attitude Adjustment (as told to me by Carl Haglund)

In the late '60s Inco had a base camp near Savant Lake with temporary but comfortable conditions for area supervisor and crews, drill supervisor, helicopter pilot and mechanic, and with a core shack and kitchen/dining room.

The helicopter mechanic was young, mouthy, and soon proved to be a major pain in the butt in camp, always talking about his prowess with the opposite sex. No girlfriend or wife was exempt from his claim that he had bagged her, or could do so if he had the chance.

The men tried to get along with a minimum of friction, but the mechanic continually provoked people, and resentment was building. Carl was afraid that someone would eventually take a round out of the young fellow, which could lead to people getting fired, an unnecessary complication in any exploration camp, so Carl devised a plan.

From time to time a flight to Sioux Lookout was needed to record claims, or for doctor visits or whatever. There was usually room in the aircraft for extra personnel to go shopping or just have some R&R. A flight was coming up and two or three men and the mechanic were going along. The mechanic was bragging as usual that no woman in town would be safe if he was around.

There was a lady in town who was known to liberally share her charms. Carl told one of the other men to line the mechanic up with D____, which he did.

At the supper table a few days later, Carl casually mentioned that D____ had been checked by the Sioux Lookout Health Unit and that she had VD. (bogus, of course.) The subject of VD was discussed, with Carl offering his knowledge (also bogus,) of how you could tell if you had contracted the disease.

"One symptom," he said, "Is that if you cut yourself, the wound, however slight, will not heal. Another symptom is frequent trips to the little brown shack."

The mechanic was a notorious cookie monster. When the cook made a batch of goodies the mechanic would always grab more than his share and stash them in his tent. Carl had the cook add Ex-Lax to a bunch of brownies and told everyone else not to eat any. Of course, Cookie Monster scarfed the whole batch.

Carl had also noticed that the mechanic had barked a knuckle pulling wrenches on the chopper. Human nature being what it is, the mechanic, on hearing that a cut would be slow to heal if you had VD, kept picking at the scab. You can see the poor bugger never stood a chance.

For a few days the mechanic became very quiet. One day he got Carl alone and asked if he could go to Sioux Lookout on the next trip. Carl asked him why.

The mechanic said, "I think I caught a dose from D____."

"What makes you think that?" Carl said.

The mechanic replied, "I did it with D____ when I was in town, and now I have a cut that won't heal and I'm continually going to the can."

Carl said, "We keep condoms in the first aid kit for when you guys go to town."

"I know," said the mechanic with tears in his eyes, "But it was my first time, and it fell off." By this time he was crying.

Carl said, "Look, you can't go out tomorrow because there is a load of drill core, but I'll check with the Health Unit, and we'll make an extra trip if necessary."

When Carl got back he took the young fellow aside and told him that the Health Unit had double-checked, and that D____ did not, in fact, have VD.

Carl took pity on the young fella, and never spoke a word about his tearful confession.

The mechanic, to his credit, became a changed man, and finished out the season as a team player.

Ten years later that same guy flew a one-armed goose-hunting Beaver pilot from Shebandowan to Thunder Bay – no doubt saving a life.

Now that is an Attitude Adjustment.

Table Scrap: Well-done Steak (as told to me by Carl Haglund)

The base camp had moved on from Savant, but Carl still had some drills in the area serviced by float plane. He had rented a small house near the bar to operate from.

On this particular summer day Carl had decided to repay some local hospitality by having a barbecue, and to this end had procured a good-sized grille and purchased some nice T-bones.

Everyone in Savant had a dog, it seemed, and some of them would wander in small packs and get into mischief, as dogs will do, just like kids. To keep the steaks out of the dogs' reach Carl had put the barbecue in the back of his pickup truck. He got a good bed of briquettes going, put on the steaks and retired to the bar, sitting near a window where he could monitor the cooking process while enjoying a bubbly.

Someone popped in and said that they had heard a plane come in. Carl wasn't expecting any of his, but figured he'd better check. Out to his truck and down to the lake he went, forgetting he had steaks cooking in the back.

To get to the lake he had to go a quarter of a mile down to the rail crossing, a quarter mile back and then a quarter mile down to the lake, about a two mile round trip.

He got to the lake, saw that it wasn't his plane, turned around and went back. The road was smooth and somehow the forgotten grill did not capsize. Someone told him later that it was quite a sight, with the smoke from the well-fanned fire streaming from the back of the truck.

As he got out of the truck he said, "Holy cow! I forgot about the steaks!"

Well, you might say they were well-done – a collection of bones with pieces of charred meat hanging here and there.

The local dogs got their feast after all.

Chapter VIII Summer '71 – Spring '72 (Treading Water)

I'm mostly working alone now, picking up casual help if the need arises. Inco gives me two staking contracts totalling 40 claims (my reward for ratting out the airborne turncoat?) The claims are between Dryden and Sioux Lookout and straddle a forestry access road. Piece of cake – I work alone and make a buck.

An ex-Inco associate wants a mag survey done on two small blocks on the shoreline on the west arm of Red Lake. The lines are already cut and chained. The jobs are less than 10 line miles each and will require a camp move, so I figure I can save time and show a profit if I hop in everyday by floatplane.

The first day, on my first loop, I run into trouble. A bartender had been hired to do the cutting and chaining, and he should have never quit his night job. The lines are well cut and straight enough, but the contractor was obviously unclear on the grid concept. I run my first line north and hit the claim boundary at 14+60N (1460 feet) – so far so good. I cross over to pick up the next line and find to my surprise that the first picket is marked with a zero. I go down the line to the next station. Just as I feared it is designated 1-S (one hundred feet south.)

Here's the deal: Mr. Bartender should have started chaining all his lines at the baseline, but he was too lazy to walk back each time I guess, so he just moved over and started back south. Even that might have not worked out too bad if he had started at 14+60 and worked backward. I read the darn line anyway and when I reach the baseline I am at 14 south according to dipsy-doodle line cutter (I share a few other adjectives with the whiskey jacks, all unprintable.) I consider my alternatives; I can't rechain the lines myself, and reading the bogus back lines will create confusion on the map. I decide to read every second line and leave the others, and by the time the 180 comes in I am done (half-done, really.)

The next day I fly into grid #2 and find the same situation, so once again I read every second line and fly out. I return home and the next day I call the ex-Inco guy to complain and explain. He is strangely unperturbed. I tell him I will send him the readings and he can generate his own mag maps. He agrees to pay me for work done, but I clear less than 100 bucks; ½ step ahead, ¼ step back.

(Years later I discuss the deal with a mutual friend. He tells me that the ex-Inco guy had become a semi-shifty promoter after leaving Inco and things finally locked into place in my feeble brain. Ex-Inco didn't give a rat's ass about the quality of the work. He just wanted to be able to report that an instrument survey was underway on 2 grids blah blah blah: thus mining a few more widows' purses for the gold therein.)

In August I land a day/wage job with Kerr Addison, another division of Noranda. They maintain a field office in Kenora, and I have become acquainted with Jim Campbell, the head honcho, during the last year or so. He wants me to do an EM/mag survey on a lead/zinc option near Marathon. Oh no! Back to the north shore of Lake Superior.

It's nice to see Heron Bay and K.T. McCuaig again. I have Ralph O'Grady and Allan Donnely, two of Jim's men with me and we overnight at the Heron Bay Hotel. Tomorrow we will chopper into a small lake two miles east of the Pic River where a camp awaits.

The next morning a Bell Jet Ranger arrives and ferries us, our groceries, and our geophysical stuff into the bush. This is my first experience with the Jet Ranger and I'm impressed. It's strong, roomy and fast and once again the pilot is excellent. In no time at all we are in camp and the chopper returns to Rossport, less than an hour away at Jet Ranger speed. He will be in and out semi-regularly, as Jim has a Winkie crew drilling on the showing. We have no radio, but in case of emergency we can walk to the Pic River where a canoe is stashed.

Table Scrap: I sell the morning paper, sir – Shield Style

The Heron Bay CPR Station had been torn down leaving an expanded parking lot between KT's, the hotel, and the CPR mainline. This was where we loaded the chopper on our way into the bush.

KT sold a morning paper – probably the red-eye edition of a Toronto paper. The paper bundle was delivered by the westbound Super Transcontinental, and that rascal went through Heron Bay like a whirlwind.

We were greeting the chopper. He was just settling down when the Super-Connie whizzed by. The newspaper bundle came in like a missile from the open baggage car door – high, hard and inside, like a Juan Marichal fastball, missing the still-spinning tail rotor by inches. If the strong-armed paper boy had delivered single issues I'm sure they would have landed in the eavestrough on the second storey of the Heron Bay Hotel.

The Jet Ranger pilot was <u>not</u> pleased. Subsequent chopper landings were further from the tracks.

Most of the grid had been cut earlier in the summer. We will do the EM and mag work on 12 line miles, adding a 4-mile extension. Ralph runs the mag – Alan and I will do the geophysics using the new CEM.

This is a shake-down cruise for the CEM. Crone geophysics has designed and manufactured this unit as a second-generation follow-up to the trusty JEM. The CEM. works on the same shoot back theory, but is extolled as a lighter, stronger unit with more depth penetration. One interesting feature is the fact that cumbersome and uncomfortable earphones are unnecessary – a field-strength meter replaces the audio. Jim is rather excited by the

CEM's potential and is rightly proud of the fact that we are one of the first to use it. I'm not from Missouri, but you have to show me. I reserve judgement.

Right from day one it is clear that the CEM is a pretty touchy unit. This is the Superior North Shore, remember, and the weather changes more often than a politician's ethics – clear in the morning, low overcast by noon and vise-versa. Hell, some days its versa-vise and back again. I am having trouble getting consistent readings out of the CEM, and we read and re-read lines trying to find some base parameters to give us some reasonable expectations that our survey will be accurate. To further complicate things, we are dealing with typical North Shore terrain with heart breaking hills and nasty under brush. We read a line once, twice and yet again, and my frustration is mounting.

It takes a while, but I finally figure it out. The CEM is super-sensitive to atmospheric conditions. As soon as clouds appear the field strength meter goes goofy, and clouds appear every day at one time or another. We have to pick our survey times carefully. Seldom do we work more than three hours. When clouds appear we lay the CEM down and go to the other end of the grid to cut some picket lines – willy-nilly work for sure.

Ralph has finished the mag work. The lines are cut and chained and Alan and I move on to the fresh-cut lines to finish the CEM survey, only to find that this machine has been holding back on one more eccentricity. The four-mile extension covers the only flat terrain on the option, and we hope to bang the last four miles off and head home – Mr. CEM has his own idea on that.

It turns out that it is also overly sensitive to conductive overburden. We find conductors everywhere and they keep changing location. I switch over to vertical loop mode and find I can move conductors around at will. Three days later we give up. The survey is useless.

The next day the chopper pulls us out and we head home to Kenora where I give my report on the CEM to Jim. It's taken us almost a month to do two weeks' work and half of that is of no good to Kerr Addison. I feel bad about the whole deal, but Jim is sympathetic. He sends the CEM back to Crone with my critique and (although I'm not sure about this) the CEM project is scrapped. I return home to lick my wounds once more.

Table Scrap

The CEM was certainly a pretty unit to look at. Both transceivers were finished in bright yellow bakelite and they looked pretty darn impressive if you ignored the evil lurking within. One day Alan and I were sitting on a log staring at the two wheels and their hidden secrets. I leaned over and using a magic marker I wrote NFG (no flippin' good) in big black letters on one of the spokes.

Before Jim had a chance to return the CEM to Toronto he was tapped to do a little talk on mining exploration to some grade-schoolers. He took the unit with him as a show and tell aid and one of the kids asked him what N.F.G. meant. Holy cow! Jim had forgotten to remove my field assessment and he had to think fast. "Umm, err, that means non-ferrous gradient." Nice save!

Table Scrap

Ralph O'Grady was another graduate of Haileybury School of Mining Technology. This Marathon job was his last for Kerr Addison – he was due to leave for a European backpack tour, and was pulled out a few days before Alan and I. He had cut the last line of the four mile grid and when Alan and I reached the last station Ralph had made a <u>huge</u> blaze on a <u>huge</u> birch. On the blaze was a picture of a backpacker with a meteorite arcing overhead. "Gone to follow my star" was written underneath. We should have had a camera.

Sidebar: Three or four years later, on a propane haul to Red Lake I ran into Ralph (Olympic Rings.) He was working for Campbell Red Lake on the air-movement team, his job being to test air quality underground. (This was before Campbell and Dickinson joined forces as Gold Corp.)

He told me that one day he ran into a geologist deep down in the mine and the guy, knowing Ralph had some geology background, had something to show him.

It was the pay stope, and Ralph said the face was yellow – not streaks of yellow veins – it was solid yellow! Ralph laughed – iron pyrite? Nope – it was gold, and a sight to behold!

Back in the day (and maybe still) most gold mines had a pay stope. They were often blocked by heavy doors and locked! When the mine needed to boost the grade at the mill head, the "vault" would be opened and a round of high grade would be extracted.

It's September now, and I am off to the North Shore again. Ex-Inco Guy wants 50 claims staked on the old Zama Mine road near Schreiber.

I take Rolly, a local guy, with me and we head for Schreiber in the old Pontiac. Rolly has never staked a claim - he cuts pulp for a living, but he is a good man and I have no doubt that he will catch on fast. We check into the local hotel and the next day we hit the bush and man, is it dirty!

The hills are not too bad considering it's still the North Shore, but the whole area has been clear-cut and has grown back in thick balsam fir rabbit bush. Some kind of disease has killed the balsams before they reached a height of ten feet, and as a result they are all dry with sharp dead branches, waiting to snag your clothes or poke your eye out. The first day I meet Rolly at a corner post, and he says he has never seen such miserable bush. He swears he saw a woodpecker fly by carrying a lunch bucket.

Four days later we finish and head home. I've actually made a buck this time, but it's not really my money. I have lots of bills to pay – lots of bills.

The highlight of the Zama Lake staking job is an overheard conversation in the bar of the Schreiber Hotel, and what an interesting eavesdrop it is.

At the next table four guys are chatting. One of them is a well-known Thunder Bay geologist and he is telling his pals to watch New Brunswick Uranium (NBU.) They have pulled a Bonanza hole at Ignace and the stock has gone from pennies to $1.80 already. I perk my ears up right smartly.

You will recall the sliver of ground staked by Jack, Garth and Fast Freddy during the Mattagami Lake rush, and the fact that they had optioned the claims to NBU. I am intrigued, and when I return home I certainly do track that stock. The Thunder Bay guy was right on the mark. NBU starts to climb and that sucker goes up a dollar a day for 30 days, peaking out at over 30 bucks. Holy cow! I wish I had kept my Seemar cash. I could have jumped in for 1000 shares and made 30 grand. Oh well – it's no use crying over spilt milk, but darn it all anyway! I would have liked a sip before the glass tipped.

I'm sure happy for Jack. Although I won't see him again until 1990 I keep an eye on developments and I have a bit of knowledge about the deal Jack and his partners have with NBU. In addition to cash and stock they have a 3% net smelter return and when NBU later goes into production, I do the math. The ore body was mined out in 10 or 12 years but for much of that time Jack and the boys shared $3000 a day on the N.S.R. Not too shabby huh?

The rest of the fall until November is pretty darn quiet. I take a crack at couple of Winkie holes; one for Ex-Inco Guy at Red Lake, and one for Kerr Addison Jim on Lake of the Woods. Both are unsuccessful. Neither one of them own a 10 to 1 reduction gear, and both holes are spotted in obvious boulder beds. With no reduction unit I jostle and bounce my way through boulders, chewing up clutches and casing shoe diamonds, until I give up in disgust. Why these guys won't spend 500 bucks to save thousands and produce some core is a mystery to me. Why didn't I buy a gear and rent it to these cheap you-know-what's? To tell you the truth, I never thought of it.

I'm not un-busy though. A few years ago the Ontario Provincial Government had fired up an educational institution to provide job-related courses for the chronically underemployed. For reasons which will become clear later on, this outfit will be nameless. We will call it Empire Builders Anonymous University (EBAU.)

They have a branch office in Kenora and are teaching various courses at various local venues. The word has gone out that they want someone to design, set up and teach a pilot course in staking, line cutting and geophysics. Kerr Addison Jim and Harry Bell, the Kenora mining recorder, urge me to apply for the position and I do so.

I meet with EBAU and get the nitty-gritty. They anticipate 15 students and they want to see my idea of a course outline. This is a piece of cake for me. I have been trained by some of the best and have done my share of field training over the last ten years.

I whip up a dandy proposal, motor back to Kenora and drop it on the dean's desk. This guy looks like the Pillsbury Dough Boy, and is a typical over-educated bureaucrat. He picks up my proposal with his fingertips, studies it with an air of disdain and says he will get back to me. I guess he must think I have cooties, because as I get up to leave he ignores my outstretched hand and I feel a bit foolish. This is not a guy I can warm up to. Oh well, I've wasted time in a lot of less interesting ways before this, haven't I?

Less than a week later, much to my surprise, I get a phone call from EBAU. Out of 50 applicants I have made the top ten short list and they want another interview. This time things go better, probably because Frank Cornell, the course coordinator, is sitting in and he is an ok guy.

Two hours later I am left alone for a bit while the dean and Frank have a tete a tete. They return to tell me I've got the job! Super! I sign a contract for four months work at $1000 per month. The course itself will run from Jan. 10th to April 15th and I am pretty excited about the whole deal. Not only do I have an assured decent income for the winter, but I also buy in whole-heartedly to the idea of graduating students with a good grasp of the exploration concept. The student body will be predominately native and I have no problem with that at all. I have fond memories of 1960 and how patient Charlie Mcleod Jr. was with a green wet-behind-the-ears white guy. Charlie will be my role model.

140

Now, don't think that I'm strutting around like a peacock – I know why I got the job. For sure I had drawn up a good course outline, but I know darn well other applicants would have been equally qualified. In fact, included in the short list was Carl Branch, my party chief on Nighthawk Lake in 1965, and Carl knows where the Whiskey Jacks stash their breadcrumbs.

My strength was in location, location, location. My proposal was to use the 1968 Bridges Twp. property as a training area. It had all the requirements necessary - road access, a legitimate prospect with an established conductor, and it would cost less than a remote camp, and, let's face it, I had Harry Bell and Kerr Addison Jim on my side.

My first duty is a bit disconcerting, though. I am to work up a performance objective chart, and they give me a desk to work on. This is my first experience with useless paper work, and much to my dismay I find I am good at it. Three hours later I hand Frank a dozen pages of pure bafflegab and return home. I have a week's worth of (unpaid) EBAU work to do before leaving for Flin Flon. My phone line will be hot.

Flin Flon

Kerr Addison Jim has a large claim block on Winisk (Owl) Lake, thirty miles or so west of Flin Flon, Manitoba and just inside the Saskatchewan border. Half of the property is on land and the dry-ground work was completed last summer. Jim wants me to extend the grid onto the lake and complete the EM and mag survey.

No problem. I am re-invigorated now, and with an assured winter's employment ahead of me my attitude has certainly been adjusted. There should be no problem clearing this up before my EBAU contract starts.

On November 5th I leave Kenora with three men (one is Alan Donnelly) and head to Flin Flon in Kerr Addison's GMC panel van. Yes – another GMC. The Noranda organization must be in love with these slab sided, noisy tin cans. We don't have a huge load – survey gear, personal stuff, and a single track 12-horse Skidoo. Three of us sit in the front and we make a comfortable nest of sleeping bags beside the Skidoo for the fourth man. It's a long haul. We eat supper at the Pas and arrive in Flin Flon pretty late. Rooms have been booked at the Royal Hotel and by 11 pm we are glad to hit the sack.

We have an early breakfast and plan our day. First things first, and first of all we have to do a tour of the town.

I've always wanted to visit Flin Flon – who hasn't? The name alone conjures up images of grizzled prospectors in mackinaw pants and high-laced leather boots. Throw in a couple of dog teams and some snowshoes, (canoes and black flies in the summer months,) add some bannock and beans, and you have my imaginary Flin Flon.

But this is 1971. The miners are mining and the smelter is smelting. The wide main street is lined with flourishing business establishments and new cars are everywhere, not a rattle-trap to be seen. We are far from being disappointed, though. This is a frontier town, and on the frontier it's the people who make the difference. Seldom do we pass anyone on the sidewalk without receiving a nod and a smile. Also, Jim and Kerr Addison have been visible on the Flin Flon radar for many years, and we are instantly accepted by association. A red carpet keeps rolling out ahead of us. Pretty darn cool, eh?

Now we hit the grocery store and once again I am in Timmins, Red Lake, and at K.T. McCuaig's. They must clone these guys, I'm sure! We don't have to do much shopping. We just tell the manager we need grub for four men for ten days, and that's about it. The order will be ready tomorrow morning and we will re-supply by aircraft on a weekly basis.

We run into a snag in the ointment. It looks like we can't get to the camp right now. Our plan A was to drive to Sherridon, a small mining community 20 miles west, continuing past Sherridon a couple of miles to a collection of homes on the southeast shore of Winisk, and on by Skidoo 15 or 20 miles north on the ice to the camp, where two wood frame 12x14 tents await our arrival. We had expected to find freeze-up well advanced by this time at this latitude, but Mother Nature isn't co-operating. It has been warmer here than at Kenora and although Winisk had frozen over the night before, the ice is not thick enough to chance a Skidoo trip. We have to go to plan B.

So, I motor down to Parsons Airways – yep, the same Parsons we used for Squaw Lake at Kenora. They are the charter of choice in Flin Flon and have a good assortment of aircraft here. I explain my problem and we find a solution. They have a 180 on floats and their lake is still open, but not for long. Cold weather is forecast, and the Cessna is going into the hanger tomorrow afternoon for the changeover to skis. Yesterday the 180 flew over our camp on a return trip, and although Winisk is locked in, there is a small lake a mile away that is still open. They agree to put me in to the small lake tomorrow morning before the 180 goes to the hanger. I talk it over with the boys and we decide on a modified plan B. Alan and I will go in with a minimal amount of grub and open the camp. Kevin and Don will join us in a day or so and we hope Mother Nature will co-operate.

The next morning the 180 drops us at the south end of the unfrozen lake where a half mile portage leads to Winisk. At the other end of the portage we will cross a mile-wide bay of fresh ice to the campsite – and we hope the ice will hold us.

The flight is a routine 45-minute hop, but I can remember every minute. Our pilot is a girl, and a very beautiful girl at that. Alan and I are tongue-tied. It's tough to be debonair in a stinky old parka, and Febreeze has not yet been invented. No matter – the cockpit of the 180 has a pleasant air of perfume and Alan and I are both thankful that we passed on the pork and beans last night. Yet even in our haze of hormonal fantasies, we agree on one thing. This lady is a dam good pilot.

After deplaning we strap on our packsacks of groceries, lash our sleeping bags on top of the load and carry our snowshoes across the portage. There isn't enough snow to require the shoes, but we need them to cross the ice. The snowshoes will distribute our weight and we need all the help we can get.

Before we cross the bay to camp we cut ourselves a couple of skinny 15-foot poles to carry. I don't know why. If we sink into 40-degree water they aren't going to do much to delay the inevitable. It's kind of like taking Vitamin C in flu season – you might still catch the flu but everyone says "eat pills" so you do.

We don't hug the shoreline. There could be a chance of thinner ice along the sheltered shore, so we head straight across. Mother Nature has done one favour for us. There is always a chance of snow before ice, and slush does not set up as well as water. This year the ice came first, and the first pure, dark blue ice is very strong, with no pressure cracks or ridges – it is just one unending sheet of dark luminescence. One thing we do, however, is to keep 100 feet apart and Alan doesn't follow in my footsteps. He bird-dogs me off to my left and a little bit behind. He has his instructions - if I go through he can't help me. He is to turn around, boogie back to shore and carve me a headstone.

Halfway to camp, for some reason I decide to check the ice thickness. I don't even give my axe much of a swing, I just more or less drop it on the ice ahead of me and holy cow! The blade goes right through! There is less two inches of ice, we are in the middle of the bay, and I don't do that again. I hold my breath, motor on and I don't look back.

When we reach the far shore we <u>do</u> look back, and our trail is visible all the way back to the portage. Every snowshoe step is now marked by a small pool of water on the ice surface!

Why did we chance it? Your guess is as good as mine. Probably it is because we have a job to do and we don't have much time to do it – Christmas is coming. But also, any good bushwhacker knows that the first ice of the year, although thin, is the safest. It is laid directly on the surface of the water, with no air pockets, no cracks. and has not yet become brittle. Two inches of new, unblemished ice is safer than four inches of the cracked and brittle variety.

We find the camp in remarkably good condition. We remove stored equipment, sweep out a few cobwebs and spend the rest of the day cutting firewood. We hope the other two guys will join us in two days.

That night is clear and cold. The temperature drops to -40 degrees and Mother Nature goes to work.

I'll tell you something right now, and you'd better listen. If you ever get a chance to sit on a quiet northern lakeshore while she's making ice, you don't want to pass it up. The pressure of the forming ice creates cracks on the lake surface, and on a cold, still night it sounds like whales talking. At 9 pm Alan and I step outside to listen and marvel. A crack starts in our bay and yodels across five miles to the east shore, fading off into the distance, only to be followed by the faint sound of another coming up from the south, seeming to pick up speed and volume as it heads north past us and fades off into the distance. Faint songs and loud songs, some ending with a loud crack like an explosion. And overhead, as always, the Northern Lights are dancing. Winisk Lake sang us to sleep that night.

About 2 am we wake up to one hell of a ruckus from back on the portage. It sounds like a party is in full swing with howls, yelps and yaps. We decide it must be a family of foxes and we go back to sleep.

The whole dam country froze that night. In the morning I go down to fetch a pail of water and I have to cut through more than six inches of ice, and Mother Nature continues to build on it for days.

That afternoon a 180 on skis brings in Kevin and Donny and the rest of our grub. The pilot (not the beautiful lady unfortunately) says that there is signs of a fresh moose kill at the other end of the portage. My neck hairs stand up. Alan and I are glad the wolves waited for the main course and passed on the finger food. Man – I love this country!

We get to work right smartly. First we locate last summer's bush picket lines, check shore chainage, and check a few stations to adjust our mag to last summer's map. Tomorrow a plane will be going by and will drop off some laths. We have many lake lines to establish.

Once again Mother Nature is co-operative. The old girl follows the cold snap with a snowstorm and we have enough snow to plant our pickets. The work goes well, and by the end of the week we are pretty far from camp with our picket lines. It's time to bring in the Skidoo and sled.

Alan and I catch a backhaul on the grocery run and spend the afternoon shopping. We pick up more laths and extra gas for the Skidoo and power saw. That evening we pop into the bar and watch the band setting up. They cover the early Beatles including mop tops and skinny-leg pants. They also have a ton of equipment – the amp stacks look like tall buildings. They need them in the Royal, the Thompson Inn and the Royal Hotel have the largest bars in Manitoba, seating 400 patrons. We decide to stay for the show and we are glad we did so.

The next day we drive past Sherridon to Winisk Lake, load up the sleigh, fuel up the Skidoo, and head up the lake. The GMC is left in the corner of the government dock parking lot. No one will mess with it, not in the North Country.

We are a little more than five miles up the lake when the Skidoo runs out of gas. What the heck? We filled it before we left. It doesn't take long to figure out the problem. The steel gas tank is built into the belly pan beneath the engine and is really part of the frame. This is an old girl, and stress has broken a weld along the upper edge of the tank. Hmm – so that's why we can smell gas as we travel along.

We have no time to return to Flin Flon for a major repair job, so we figure out a bushwhacker solution. We half fill the gas tank, stuff an oily rag in the crack to keep the snow out, and by leaning the sled to the right we can get ten miles out of a refill, and that's how we travel for the next three weeks. Works out pretty well too, as long as we stick to right hand turns. We'd never make it on the Nascar Circuit.

Sidebar: I love that old TV ad with Nascar's Jeff Burton. He's teaching his kid how to drive. "Turn left," says Jeff, and further on "Turn left."

The kids asks him why not turn right once in a while?

Says Jeff, "Why would anyone want to do that?"

December is cold and windy. For two days all the air in North America rushes up Winisk towards the Arctic, rests for a day and spends the next two days rushing back south. Reset and repeat – over and over. It's a good thing we are near camp. Two hours on the lake requires an hour around a hot stove thawing frozen cheeks and fingers. We don't dare leave any gear on the lake overnight – we would never be able to find it the next morning. Occasionally we have to redo sections of picket lines. If the wind has not blown the pickets away, the snow banks have covered them. Our progress is slower than we expected.

We finish the JEM and mag on Dec. 22nd, but we have no time to do the vertical loop work before Christmas. Alan and I will return to finish the job before the New Year.

December 23rd I send Kevin, Donny and their bedrolls back to Flin Flon by air; Alan and I take the Skidoo to Sherridon and drive to Flin Flon. We park the van at Parsons and we all cab it to the airport to catch a Transair 737 to Winnipeg. Alan has family there to meet him. Jim takes the rest of us to Kenora, and late that evening I am basking in the warmth of hugs and kisses. Once again, my wife has had to shop for her own present.

Christmas morning we arise early to see what Santa Claus brought us and very early it is: it's pretty tough to sleep in with two little rug rats lifting my eyelids to see if anyone is in there. Christmas dinner is pretty much a brunch. I have to be at the airport in Winnipeg at 4 pm; Alan and I are flying back to Flin Flon.

The flight north is a circle route, with stops at the Pas, Thompson and Lynn Lake on the way to Flin Flon. The landing at the Pas is pretty much normal, but when we hit Thompson I start to get a bit worried.

The landing is normal – touchdown followed by reverse thrust and brake application, but I'm pretty sure I hear some strange noises beneath my feet. It sounds just like when your vehicle needs new brake shoes – metal-on- metal. When the 737 swings around to head back to the terminal we seem to be pretty close to the end of the runway. I know these northern airports have little room to spare, but this is too close for comfort.

The Lynn Lake landing is pretty scary. By now the noises beneath us are unmistakable – shrieks, clatters and bangs. It sounds like parts are falling off and at the end of the runway we make more of a power turn to the left instead of a gentle swing. We are sitting on the right side and our wing passes over stunted Labrador tea, and the tree line is not far away. Holy cow! We have used every available inch of tarmac!

At the terminal the pilot makes an announcement. We will deplane while the ground crew inspects the aircraft. No shit, Sherlock! They couldn't keep me on this hunk of tin with handcuffs! We all walk through blowing snow to the terminal.

By the time the locals pick up their loved ones, there are 15 or 20 of us left in the tiny waiting room with 10 chairs to sit on. An hour later we are getting no feedback from Transair and we are getting cranky with the ticket counter girl. She's pretty frustrated also, and finally a transportation fleet is rounded up. We go in to Lynn Lake where the hotel dining room has agreed to stay open late to feed us.

We have a decent supper but we are not out of the woods yet, not by a long shot. The dining room staff regretfully informs us that they have to lock up. We will wait in the hotel lobby, and the lobby is smaller than the terminal waiting room.

We are quite an eclectic bunch. Besides Alan and I, there are three young fellows who play junior hockey for the Flin Flon Bombers, a couple of families with young children, one a mere baby, and a grey-haired sprightly old lady. There are few places to sit and less space to stand. There is still no word from Transair, and I'm starting to wonder about our travelling companions. They seem to be like sheep in a slaughterhouse, so Super Staker swings into action.

I go to the front desk, borrow the phone and call the airport. I want to talk to Transair right now, and Transair comes the phone. It takes a lot of argument and brow beating before the truth emerges. The brakes are shot on our aircraft and a replacement is being flown in. When can we expect to continue our flight? They say one in the morning. Why? Well, the plane is on its way from Toronto – Toronto for Chrissakes! I point out that they have young children, old people and pee-poor facilities here – and it's only 9 pm – and we want rooms!! They finally agree to rent two doubles for us and I pass the phone to the desk clerk. She fills out the registration cards and I sign them, with the added notation "for Trans-Air."

So we have two rooms and as life flows, so does our mixed bag of refugees. Those of the tightly wound variety go to one room to rest and contemplate on the sinners of the world. Alan and I are joined by the Flin Flon Bombers and a couple of others in the other room. We are going to party baby, and party we do. We keep the beer vendor busy and have a great

145

time. The star of the whole show is the grey haired lady. She has spent her whole life in the North Country and she knows where the bears hang out and all their secrets. We all love her to bits.

At 2 am they pour us into taxis and take us back to the airport where the Toronto 737 awaits. It could have been a Lower Slobodian Airlines biplane for all we care. At least half of us are a pretty happy bunch.

The flight to Flin Flon is kind of funny, though. The prim and proper avoid us like the plague. They keep their distance at the back of the plane and try not to make any eye contact with those of us at the front, as we giggle, snort and belch. By 4 am Alan and I are snug in the Royal Hotel. We have to sleep fast – tomorrow we have work to do.

We bang off the vertical loop work and secure the camp, which will remain for the diamond drill follow up. Alan and I toast the New Year with a cup of coffee. New Year's day we leave for Sherridon before sunup and it's a long drive to Kenora. 1971 has gone bye-bye.

Two weeks later I receive an invoice from Transair for the two rooms. I drop them a line to decline the offer, and tell them what I think of them. Then they try Kerr Addison, and I have to stroke Jim a bit. I sure hope he never paid that bill. I promise myself I will never fly Transair again, but somehow I don't think they missed me at all, at all.

Chapter IX   1972 – Bush Addiction Recovery Program – Methadone Phase.

On January 3rd I am back in Kenora at EBAU. Before leaving for Flin Flon in November, I had spent a week on the phone with old friends and previous industry associates. Thanks to my wide range of contacts I had scored a plethora of geophysical instruments for the course. Some companies have given us a half price rental deal and Crone geophysics will supply a JEM and a VLF (very low frequency) unit free of charge. Inco, who I have criticized in Book One for their penny-pinching ways, has opened the doors to the vault – anything I want they will supply at no charge – go figure.

So I have already saved EBAU a considerable ton of money, and this week I will add to the plus side of the ledger. The college has been making arrangements with a motel in Vermilion Bay, and they will put the students up in four rooms. Meals will be supplied, and EBAU will rent a 12-passenger van, which I will use to drive the students to and from Bridges Township. I flash the stop sign – hold the base runners! And I point out a thing or two.

The motel in town has no classroom facilities, but they do have a bar. We need a classroom, but we certainly don't need a bar. The restaurant/gas bar where we had worked out of in 1968 is right across the road from the proposed instruction area. The business has been sold, and the cabins have been replaced with a new 4-unit motel. The students can live, eat and study in one place, and the grid boundary is a short walk across Hwy 17. On the way home that evening I stop and talk to the new owner. He will be happy to oblige. Hunting season is over, summer fishing season is four months away, and he has a 14x20 frame building behind the motel which will serve as a classroom. I give him the EBAU phone number and he strikes a deal.

The rest of the week I touch base with the various government agencies involved. Employment Canada will supply an income for the students. Most are already on seasonal benefits and those who aren't will be enrolled. The Dept. of Indian Affairs will outfit the group with proper clothing and classroom stuff (pens, pencils, rulers etc. etc.) The Dept. of Lands and Forests will lend us axes, bed rolls and snowshoes. We are all set – the course will kick off on January 10th.

Table Scrap

A wicked blizzard hits one afternoon and the Trans Canada is shut down, marooning me in Kenora. I get a room at the Kenricia Hotel and to my surprise, the same Beatles cover band we had enjoyed in Flin Flon are playing in the bar tonight.

I decide to take in the show and when the boys hit the stage I am sitting about 30 feet away with a full glass of beer in front of me.

Now, this is no comparison to the Royal Hotel venue – not by a long shot. The Kenricia Bar is in the basement and the ceilings are not much more than 10 feet over my head. It seats 100 at the best of times, and the group has used up 20 percent of the space with their array of amplifiers. They open with "She Loves Me" and when the first chords hit the amps, the whole room vibrates. My beer starts to foam and runs over the top of the glass! My ears are hurting! I sneak out the back door and go to my nice quiet room on the second floor – I'm too old for this.

COURSE OUTLINE

GEOPHYSICAL ASSISTANT COURSE

PILOT PROJECT

Retraining Program,
Districts of Kenora and Patricia.
January to March, 1972.

A twelve week Geophysical Assistant Course commenced January 10, 1972, at Bill's Halfway House with eight students enrolled. The objectives of the course were twofold: first, supply training for the students that would give them a better chance of employment upon graduation and secondly, to use this course as a pilot project for future courses of this type. The course was terminated at the end of the ninth week due to the small enrolment and poor attendance of four of the six remaining students. Only two students completed the nine weeks and were offered employment on geophysical exploration crews the following week. Information and experience obtained in the planning, administration and operation of this type of course will be valuable for future programs.

R. R. Durnin,
Instructor.

COURSE LAYOUT

WEEK ONE

One hour each day or one full day devoted to first-aid training. St. John's Ambulance usually have instructors available.

Two hours devoted to ski-doo, outboard and power saw operations and repairs. Local ski-doo retailer will lecture.

Two hours local Lands and Forests Branch could supply lecturer on bush survival, general bush safety, fire season, camp requirements and safety, and safe canoe handling. Also, camp clean-up requirements and work permit regulations.

One to two hours radio procedures, different aerial set-ups. Department of Transport representative could lecture on this.

Remainder of the week devoted to camping - selection of site, methods of construction, both fly camp and permanent. Plywood and 2 x 4 frame set-up, santiation requirements, equipment and food check lists, projection of fuel requirements, durability and preserving of vegetables. Camp equipment requirements, summer versus winter, and care and maintainance of same. Snowshoe care, selection, different harnesses. Axe selection, care and filing. Cover care and safety re naptha stoves and lamps. Local propane outlet will lecture on use, care and maintainance of propane stoves, heaters and refrigerators and lights. Possibly some oil stove instruction.

WEEKS TWO AND THREE

Continuation of week one if necessary.

Introduction to field work with emphasis on staking and line cutting basics, staking laws and methods. Harry Bell, Mining Recorder, will lecture on this. Compass explanation and practice. Introduction to maps as outlined in the proposal. Suggest complete set of maps for one area and inter-relationship of said maps shown. Some elements of drafting could be covered but this would more logically follow the field work.

Line cutting would include basic explanation, grid systems, turning off lines, compass, prism, board and transit and different applications. Problems encountered re topography, offset base lines, tie-lines, correction, slope chainage and breaking chain. (Local Ontario Land Surveyor could be approached to lecture on basic transit work, time permitting). Compass and pace grid explained and practiced.

WEEKS FOUR, FIVE AND SIX

First field segment.

Practical application of the basic classroom work. Claim block would be staked and lines cut on same. Record claims.

Four students showing the most leadership qualities chosen as group leaders, these groups to work together in the field when applicable. These groups also begin camping out on property. One week each. This camping program to be adjusted to fit field work segments. As final staking practice, these groups will lay out, stake and record four blocks and two claims each.

WEEK SEVEN

Second classroom segment.

Instructor explains basic electromagnetic theory and basics of operation of different systems. Introduction to plotting, interpretation and note-taking. Some practice reading near camp. Final drafting of line-cutting performed on first grid.

Instructor will prepare outline of geophysical theory and operation to be taught. This proposal will be submitted to the District Geologist (Leo King) for approval and suggestions.

WEEKS EIGHT, NINE AND TEN

Second field segment.

Application of geophysics as covered in second classroom segment.

Reconnaissance and detail geophysical survey done. Reconnaissance and detail field mapping practiced. Field interpretation applied.

Magnotometer surveys to be practiced first. Weeks 7,8 and 9 and 10 will be flexible regarding classroom to field application due to proximity of field area to the classroom.

Emphasis on Crone J.E.M., Vertical Loop, Radem with horizontal loop, Turam and Scintilometer scheduled if time permits.

WEEKS ELEVEN AND TWELVE

Final classroom segment.

Review of previous weeks.

Final drafting and preparation of work for assessment purposes.

Leo King would give final lecture, expanding previous theory and giving examples of his evaluation of results and where he would spot bore-holes.

Leo King, Resident Geologist, could give a lecture on basic geology. Recognition of rock types and basic minerals, and explanation of some magnetic and electromagnetic anomalies through geology.

Jim Campbell could give lectures on soil sampling and testing. Perhaps some practice near base camp.

Trenching techniques explained and perhaps C.I.L. representative could lecture on the use, care and safety regarding dynamite.

REMARKS

I have laid out this course proposal with a number of specifics in mind.

Location of base camp - Bill's Halfway House, 17 miles west of Vermilion Bay on Highway 17. This camp is quite accessible to both Kenora and Vermilion Bay and Dryden for supplies, and what local lecture assistance may be required.

I have tried to schedule the course in logical progression of priorities and sophistication. This way, the student begins with real elementary work and as the course progresses to the more mentally taxing portion, he should be able to assimilate facts easier.

The course is also flexible in its layout. Should one aspect take more or less time to complete than anticipated, it can run over or under the time allotted without throwing the scheduling out.

The course includes outside help on all facets of the program where I feel I am less than qualified. In the case of First Aid, Radio Communications, Ski-doo maintainance, etc., I would be passing some of my own bad habits on to the student.

The course should stress good work habits, responsible use of company equipment and each applicant's attitudes around camp should be noted.

(See appendix for other useless EBAU crap)

The first two weeks of the course are relatively uneventful, although I do have one eye-opening and discomfiting occurrence. I have included a first aid course in my proposal, and as luck would have it St. Johns Ambulance is running an evening course once a week in Vermilion Bay. I enrol my students, and the first night I am confronted with my first episode of overt racism. There are about 15 locals taking the course, and they want nothing to do with my group. We form a little island at the back of the classroom, and when it comes to partnership exercises we work within our own group. All my life I have taken people at face value, and I'm a bit pissed at my neighbours' attitude. I can't do much about it, yet I feel bad for the boys.

Pay day! And I spring the boys loose a little early on Friday so they can catch the bus to Kenora; I expect them back on the 9 am bus on Monday. Monday I am short a student, but he shows up Tuesday. No problem, but the course has sprung its first leak and I don't have a bucket.

The students' dress code starts to change. By February half of the parkas are replaced by jackets, and the headwear of choice is a modified triangular bandage (supplied by the first aid course – there's not much of a beer parlor market for triangular bandages.) Snowshoes, bed rolls and axes survive the migration. I keep a pretty good eye on the payday bus, and besides, no one would dare to walk into a Kenora bar carrying an axe.

At the end of January I give EBAU an invoice for my daily mileage and receive a lesson in economics 101. They won't pay up; they take the stand that everyone has to drive to work. I point out that I could have walked the two blocks to the motel in town and would have driven the company van, but they are adamant. I'm pretty disappointed at the Pillsbury Dough Boy's attitude, but I bite the bullet. I am committed to the course. I want it to be successful.

Due to the proximity of the grid and the flexibility of the course I have been using the classroom on bad weather days, and on some cold mornings I spend two hours lecturing before heading to the bush. One morning I wrap up the lecture at 10 am, don my parka and no one moves. I ask what the problem is and look at Gary Wahpay, who has been sort of a spokesman for the group. He tells me they are on strike. On strike! Why? Gary says they are being treated unfairly. All the other Indians are living at the Holiday Inn and studying book keeping and accounting. Gary and his union members are stuck 30 miles away and working outside. They want danger pay and an isolation allowance!

I can't help it, and I laugh so hard I have to sit down. "Dream on," I say, and pick up my snowshoes. When I reach the grid I look back, and my rag tag bunch of miscreants are following – the wobble is over.

154

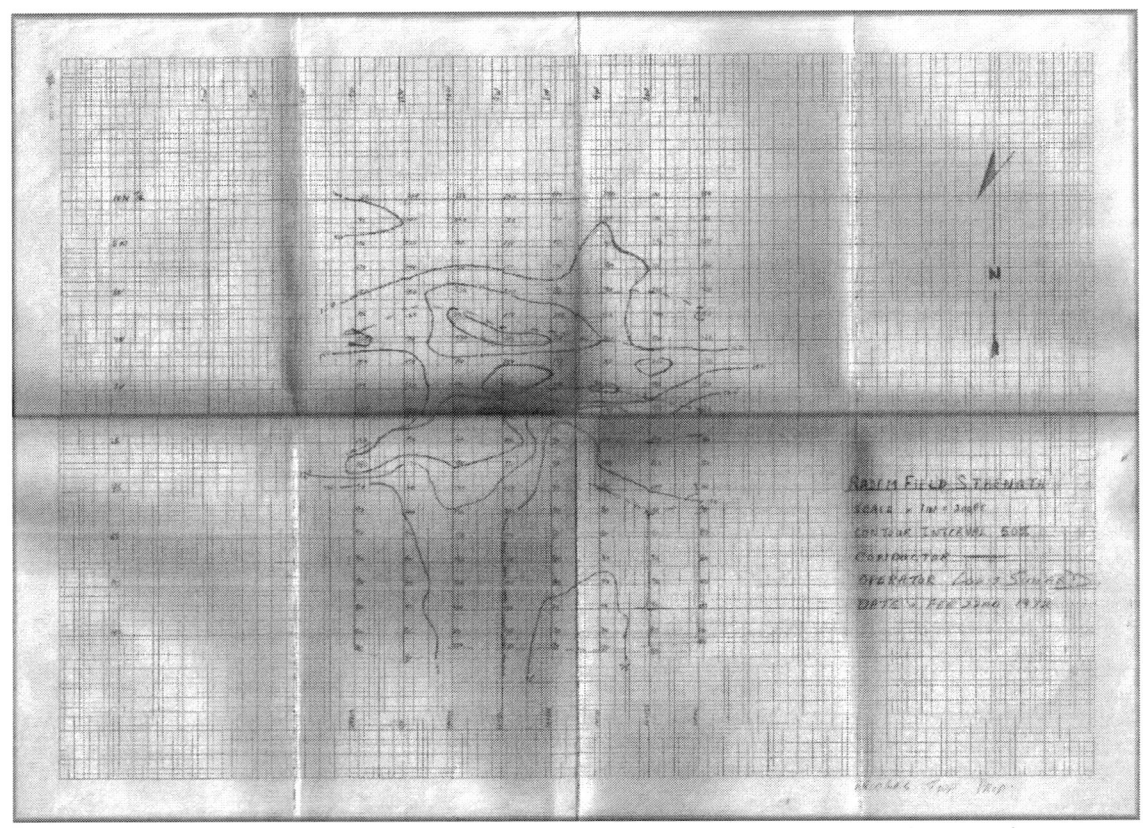

A contoured VLF map created by a student in the classroom – Nice Work!

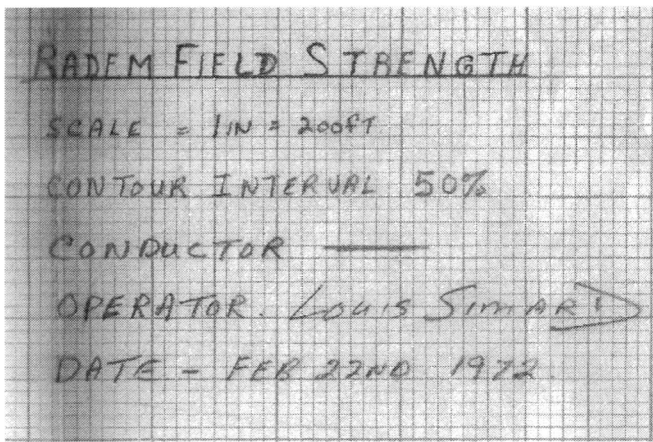

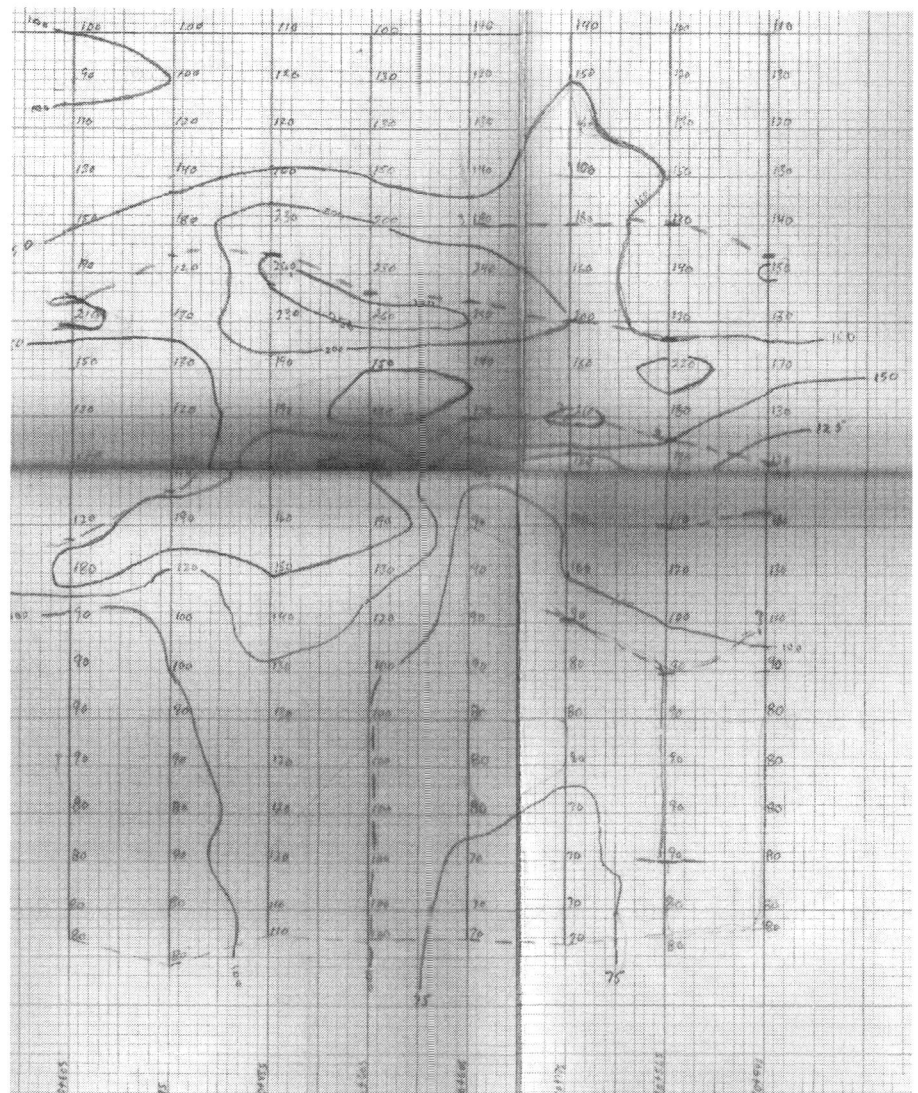

Enlarged section of the same map

The situation with Gary bothers me. He is a good-looking young fellow; always well dressed, well-spoken and very intelligent. He is a hard worker and would fit into any exploration camp – yet he seems disinterested.

The next day I corner Gary for a bit of one on one. I ask him why he signed up for the course in the first place.

"I had no choice," says Gary. "I spend my summers guiding Americans on Lake of the Woods and they are very generous. I spend my winter months relaxing and collecting pogey. One day last fall three suits showed up on the reserve and said, "You, you and you are taking this course. If you don't, your benefits will be cut off."

You could have picked me off the floor with a dust pan! I check with the others, and the story is the same. Sign up or suffer. EBAU has stocked the ship with a press gang, for crying out loud!

The next day I give the boys their afternoon assignments and boogie it in to Kenora. I have some detective work to do. I don't trot around hammering on desks; I just want some answers. The purpose of the course, as I was led to believe, is to create employment opportunities for the students. I want to find out what procedures, if any, exist to fulfill that mandate. The results of my investigation is a shocker, to say the least.

I have a chat with each of the three main agencies involved; Indian Affairs, Employment Canada, and the Pillsbury Dough Boy, and the answer is clear. The only jobs they are interested in creating are their own. Someone is receiving an education here all right, and that someone is yours truly!

The course continues to lose students and I can't do a thing about it. Some of my guys run into trouble with the law and end up in the slammer. These are minor offences, i.e. public drunkenness or vagrancy, and given their upbringing and lifestyle, is a normal hazard for them. If EBAU had any balls and a minimum concern for the boys, they could have made an appearance in court and the 10-14 days sentences would be suspended; but EBAU doesn't care, and I am disgusted.

We are down to two students, and on March 10th the course shuts down. I pack up the gear and Frank Cornell takes everything back to Kenora. He will make sure the borrowed and leased equipment gets back home.

The next day I trot in to see the Pillsbury Doughboy. He offers his sicere regrets and tells me that he needs a final report after which my employment will be terminated. I point out that I have a signed contract and he sucks on a lemon. I am told that EBAU has a warehouse, and if I insist on finishing out the contract I can shovel snow and sweep floors for a month. I tell him where he can store his shovels and brooms.

He wants me to write the report at the EBAU office. I decline the offer. I will write it at home.

That evening after supper I sit down at my kitchen table and write <u>two</u> reports. The course report is clear, concise and unjudgemental. Then I write one for the Doughboy, and it takes me a while. I point out in no uncertain terms that they have no scruples, ethics or concept of fair play. They have let my students down, and I am really, really tee'd off. At 6 am I finish my third pot of coffee and prepare my reports for mailing. I'm pretty sure I have burnt my bridges behind me, but it worries me not one bit.

Note: Official Report – page 157 – 164. Also see Appendix, page 235.

(The unofficial report exists only in my memory, and I think the Pillsbury Doughboy shredded it before he read the whole thing. Perhaps the repeated term "asshole" turned him off.)

11205

March 8, 1972.

Mr. V. Munro,
Manager,
Canada Manpower Centre,
Kenora, Ontario.

Dear Mr. Munro:

RE: Geophysical Assistant Course

This is to advise you that the
Geophysical Assistant Course will be discontinued March 10,1972,
The reason for discontinuing the program is lack of students.

On February 25, there was an enrolment
of six. The following week, two of these were in jail and one was
absent for the first three days of the week. Consideration was
given to closing the program at this time, but due to the commitment
we felt towards the students remaining on course, and the fact that
one of the students was in jail and bailed himself out to return to
the program, the course continued.

At the start of this week, February 6,
one person showed up on Monday, two were in jail, two were at a
funeral and one unaccounted for. As of today, there are two people
on course, two in jail, and two unaccounted for.

The decision to close the course was
based on the above information. The instructor has been rounding
out the two students on location.

An assessment of the course, as it
relates to the students and our learning, will be written and you
will receive a copy.

Yours very truly,

F. G. Cornell,
Assistant Co-ordinator, Retraining,
Districts of Kenora and Patricia.

FC:k

c.c. Mr. G. Clyde
    Mr. R. Durnin

GEOPHYSICAL ASSISTANT COURSE

SUMMARY

Retraining Division
Districts of Kenora and Patricia

R, Durnin,
Instructor.

159

A twelve week Geophysical Assistant Course commenced January 10, 1972, at Bill's Halfway House with eight students enrolled. The objectives of the course were twofold: first, supply training for the students that would give them a better chance of employment upon graduation and secondly, to use this course as a pilot project for future courses of this type. The course was terminated at the end of the ninth week due to the small enrolment and poor attendance of four of the six remaining students. Only two students completed the nine weeks and were offered employment on geophysical exploration crews the following week. Information and experience obtained in the planning, administration and operation of this type of course will be valuable for future programs.

R. R. Durnin,
Instructor.

COURSE PROPOSAL - BUSH CAMP

The following is a rough idea for any future geophysical courses based on a purely isolated bush set-up. If subsequent courses are to be successful, I feel this format will have to be followed. A number of objectives will be realized.

1. Problem areas A and B will be reduced by approximately 99.9%. (see Instructors Report).

2. Much better instructional facilities. Instructor living with students. No outside distractions. Living and working conditions similar to most mining company practices.

3. Course based on a 7 day work week. Why turn out a geophysical operator who expects week-ends off?

4. Student attitudes and habits develop naturally. Instead of telling the student what bush life is like, he sees it and soaks it up.

5. No alcohol problems. If the basis for a problem does not exist then the problem isn't there. A good student with a drinking problem may realize his potential and try to develop it instead of an ulcer.

6. A chance to split the course and some students into two separate categories. I think this idea is worth looking into and will be expanded later on.

The course content will approximate the pilot course, the only important changes being those necessitated by availability of instruments. However, the course will be split into two separate and equal segments.

Segment 1

- would cover course content up to and including week #5 as taught on the pilot course. Incidentally, this is being planned on the basis of five bush weeks, that is, working thirteen days, one day off, six days working and one day off. This gives us thirty-two instructional days as opposed to twenty five on the pilot course. However, there would be a heavier student load and also the extra practice would benefit the student. The course would then turn out a Class "A" linecutter and staker.

Two Week Break

-the boys go to town and blow their paycheques. At this
time, the instructor and the Co-ordinator get together and
decide which students are poor candidates for geophysics.
These are given staking and linecutting certificates and
the course is re-stacked, if necessary, from applications
received from experienced line cutters who wish to learn
geophysics.

Segment 2

Geophysical portion - run on the same instruction day set
up as Segment One. Total - thirty-two days; sufficient time
to cover the necessary work; evenings, as well as days,
utilized for class work. This gives us a total of sixty-
four instructional days as opposed to sixty on the 12 week
5 days/week basis.

Even the "shop-talk" around the evening coffee table is
instructional in nature.

Segment One would be based on fifteen students. We can
probably count on some being weeded out but even if all
fifteen completed the line cutting and staking, the student
load would not be as critical for this segment. Out of
fifteen students, there would possibly be one student to
assist. If not, then a few days assistance as proposed in
my remarks previously could be utilized. Perhaps a company
or contractor would supply this assistant.

Segment Two would have a twelve student load. An overload
from Segment One would be reduced by cutting those three
students less inclined to geophysics, and if there was a
shortage it would be made up by adding applicants. There
may even be university drop-outs interested in learning
geophysics for summer employment. They could be given a
crash course in staking and line dutting during the two
week break and be eligible for a Geophysical Program
certificate. Using this format, we would be virtually
ensured an interested, stable, student body for the final
Segment. Any attitudinal problems would be minimal.

The one big factor of this proposal is the initial cost
involved. The camp required would be a permanent tent
frame set-up; Plywood floors and walls, well built,
oilstoves, good classroom, good kitchen, fuel and propane
cache, ideally situated to serve as a bush camp for
upwards to ten years. This may seem a bit optimistic to
assume there would be a ten year requirement for this type

of camp, but I believe there is. Here is exactly what I have in mind.

AREA: Pick a greenstone belt in an isolated area, say North Spirit Lake, north of Red Lake, not too close to any settlement, and right on the Ontario Central Airlines scheduled run to Sandy Lake. Selco have gone airborne in this country; they and Noranda have worked most of the anomalies. Ask for the airborne map; if this map is kept confidential and all results are turned in to them, they should have no objection. Pick a spot for the camp so that with little travelling a number of areas can be worked. Only 15 claims need be staked for each course and six areas near camp would give sufficient rotation so that successive courses are not following fresh work done by the previous course. Incidentally, the Red Lake to Pickle Crow highway will pass fairly close to North Spirit Lake and an area could be picked that will offer cheaper transportation in the near future. Other areas along this proposed highway could be equally feasible.

CAMP: Build a good solid frame-tent camp. With proper care, these will last a long time. A caretaker would be required over freeze-up and break-up. I lived in a camp like this in Thompson that had been standing for six years and it was still in use three years later.

Requirements are:

- two double 14' x 16' tents. These are two tents built end to end to make a 14' x 32' structure. One would serve as a kitchen, food storage and cook's quarters. The other as classroom and possibly instructor's quarters.

- five 14' x 16' single tents. Four of these would accommodate sixteen men confortably. The fifth would be for equipment storage.

- associated goodies - nine oil stoves, two fridges, propane or oil cook stove, gas lamps, kitchen utensils

- cook

- cook's helper, who would also double as bull-cook, hauling water, filling lamps, oil stoves and other odd jobs. Probably not required in summer but a definite necessity in winter.

The cost of this set-up may seem staggering at first, but
average out the initial investment over a period of five years
and the cost should be quite reasonable. Kitchen wages, supplies
and fuel for the camp on a per student basis will cost less than
this winter's pilot course. Whether you eat in a restuarant or
in camp, someone has to cook. I don't think the cost per student
should be as important as the kind of student that ultimately
graduates.

There are a number of possible ways that costs can be defrayed.

1.   Instructor knows exactly how many instruments are needed
        and for how long. He reserves them for the last segment
        and rental is kept to a maximum of six weeks of utilization.

2.   Send all major companies a copy of the pilot course. Ask
        for their opinion and how they feel about course content
        and how it could be adjusted to the employment picture.
        (INCO for example have their own patented EM unit and
        their own unique field methods. However, they show great
        interest in the course and in hiring course graduates).
        Ask them what their yearly requirements are for fresh
        field personnel. Once they show great interest then ask
        for assistance, possibly an instrument and/or instructor
        loan . (Once again, we have INCO's example. They have
        the reputation in exploration as being completely
        independent as a company. Yet, they gave and offered much
        more assistance than any other company approached as far
        as this pilot project was concerned). If they are a
        company like Noranda or Kerr-Addison who may not have any
        equipment available, then perhaps a cash grant or subsidy
        for a student could be considered. These companies took a
        sharp rap in the tax mouth this year but they are usually
        very willing to assist education, especially in a field
        where they have had manpower shortages in years past.
        Even the small contractors such as Bob Penney and Dave
        Christianson at Thunder Bay would be willing to assist
        in instruction for short periods of time. A tent here,
        an oil stove there and a cash grant over here and
        expenses are cut considerably.

3.   Perhaps the labour market won't stand a possible 24
        geophysical operators every year. If care and
        consultation with district geologists are exercised, the
        chosen area could be utilized for geology and minerology
        courses. Perhaps the Ontario Department of Mines would
        use it one season for a detail mapping project. If the

industry and/or the College requires only one
geophysical course a year, then perhaps some other
educational institution could make use of the camp in
the summer. Boy Scouts, boys camps, outward-bound
projects, all could be possible tenants in the summer
season. If more than two geophysical courses are
required, then two courses could be run in the summer.
Cut the mid-term break to one week and run from May
to October.

I think the possibilities for such a course and such
a camp are endless. The groundwork and advertising
have to be methodically carried out, but once rolling,
the whole set-up should carry itself along.

I might add that there should be no labour cost for
camp construction. The first class could start a
week or so early and construct the camp. One
experienced man would be helpful.

A problem might occur - would Canada Manpower take
kindly to paying a student on the basis of 32 work
days in five weeks.

EQUIPMENT   Class equipment requirements for a 12 to 15 man
             course:

1.   Line Cutting and Staking - axes, files, snowshoes,
      compasses, etc., the same as the pilot project. More
      spares needed for bush camp. Instructor would have
      a locking cabinet to keep spare small equipment safe.

2.   Classroom - Same equipment as pilot course. A lot
      of extra paper, pencils, pens, etc. Plenty of black-
      board space and a flip chart.

3.   Geophysical- two of each instrument. This will ensure
      enough equipment to keep twelve students busy and also
      avoid undue delays in case of breakdown. Plenty of
      spare batteries should be on hand. Possibly make a deal
      with a wholesaler to return unused batteries in sealed
      boxes. I suggest the following geophysical gear be
      given consideration.

      Magnotometer - two fluxgate mags. McPhar M-700;
      Scintrex MF-1, MF-2, or MFD-2. Barringer also has
      a good digital read-out fluxgate mag. Surely two of
      the above could be borrowed. Two different types would
      be preferred for better student familiarization.

      V.L.F. EM Units (very low frequency electro magnetic)

            Crone - Radem
            Ronka - EM-16
            Scintrex - Scopex

Conventicnal EM Units

Crone JEM 480 - 1800 HZ
Crone CEM 390 - 1830 - 5010 HZ no audio signal
Scintrex SE200 or SF250 1000HZ   possibly have a newer
                                     model out also
McPhar REM or HEM 1000,   5000 HZ

Other similar instruments available. The Crone JEM lends
itself more easily to ease of instruction and is possibly
more versatile than the others. However, two different
units on different frequencys are an absolute necessity
here to avoid signal overlap when both units are on the
property at the same time.

TRANSPORTATION

If the camp is well placed, there should be no need for
motorized transportation at least for the first two courses.

In summation of this proposal, I would like to repeat:

-most problems of pilot course minimized
-a better course which would improve with time
-a need for better relations and communications with
 the public and industry
-a need for a more cosmopolitan student body; whites
 and natives - good and poor education - losers and
 winners.
-it is highly unlikely that any one instructor could
 keep a fine edge on his enthusiasm for extended
 periods of time. However, new blood brings new ideas
 and revitalization to the course. I would suggest
 that an instructor that has taught two or more courses
 be used as an assistant to break a new man in for the
 first two weeks.

ADDED REMARKS

With the College's permission, I will write to Bill Aronec
enclosing what I feel are requirements for the bush camp.
INCO has had many years experience in this type of camp
set-up and can cost it down to the penny including grub,
fuel and camp support costs. Mr. Aronec has already stated
his willingness to do this.

R. R. Durnin,
Instructor.

They say that what goes around comes around. On January 2nd 1974, much to my surprise, I receive a phone call from Frank Cornell. EBAU wants to run the course again, and unfortunately their newly chosen instructor has quit at a very inopportune time. The course is to kick off in one week and will I jump into the gap? They are willing to pay me a good dollar.

I am delighted – not at the job offer, mind you, I'm just grinning because of the fact that they have their you-know-what in the wringer. I'm not nasty about it, I had always gotten along well with Frank, but I point out that I have a permanent, well-paying job now, and why in hell would I quit, just to give EBAU the chance to shaft me again? Poor Frank is practically grovelling – they are that desperate. It's ass-covering time at EBAU and I offer a suggestion.

Do they still have the short list of potential instructors? Yes they do. Have they contacted anyone on that list? No they haven't. Man, these guys are helpless! I have to lead them around by the ears.

I suggest maybe they should call Carl Branch in Atikokan. He now owns a tourist camp, and may be interested in winter employment. Frank says he will do so.

I am talking to my wife about the deal when I suddenly realize that Carl needs a heads-up. I know EBAU and their weasely ways, so I give Carl a call. I give him a quick rundown on what to expect. EBAU has their back against the wall - don't let them off the hook too cheap.

Carl ran the same course on a remote reserve north of Pickle Lake, and with no outside influence or distractions it was a resounding success.

On the way home that spring Carl stopped in to see me. Did EBAU try to lowball him in January? Yes they did, but thanks to my warning, Carl was able to double their initial offer. I am gruntled – revenge is sweet.

Sidebar: In 2007 I see a small article in the Thunder Bay paper. EBAU plans to run a basic exploration course and are patting themselves on the back, saying it will be the first of its kind. I call EBAU and have a chat with the course co-ordinator, who sounds like a nice young fellow. I tell him that he should check some archives as my course has been run at least twice before. He is unimpressed – EBAU writes their own history.

Final note: Incidentally: through my own contacts and with no thanks whatsoever to EBAU, I secured job offers for two of my students. I was never able to follow up on their success in the exploration game, but I at least had the satisfaction of knowing that someone cared.

Pillsbury Doughboy

Please join me in remembering a great icon. Veteran Pillsbury spokesperson, The Pillsbury Doughboy, died yesterday of a severe yeast infection and complications from repeated pokes to the belly. He was 71. Doughboy was buried in a slightly greased coffin. Dozens of celebrities turned out, including Mrs. Butterworth, the California Raisins, Hungry Jack, Betty Crocker, the Hostess Twinkies, Captain Crunch and many others.

The graveside was piled high with flours as long-time friend, Aunt Jemima, delivered the eulogy, describing Doughboy as a man who "never knew how much he was kneaded."

Doughboy rose quickly in show business, but his later life was filled with many turnovers. He was not considered a very smart cookie, wasting much of his dough on half-baked schemes. Despite being a little flaky at times, even as a crusty old man, he was still considered a roll model for millions.

Toward the end it was thought he'd raise once again, but he was no tart.

Doughboy is survived by his second wife, Play Dough. They have two children and one in the oven. The funeral was held at 3:50 for about 20 minutes.

Chapter X   1972-1979 – Out of the bush! (With the occasional backslide)

The Weaning Process

Out of work again – it's time to face reality. I sit down and study on what the future holds for me and my family. On one hand I have the option of continuing to pursue my contract exploration endeavours. My alternate option, heaven forbid, is to pursue civilized employment.

I can deal with most of the bush issues: black flies, deer flies, mosquitoes, good bush, dirty bush, tent camps, isolation, deep snow, cold winters and hot summers. I can handle men, and a wide spectrum of exploration duties. What I can't handle, are the suits. On the back of my bush jacket is a sign –"Kick me." I opt for civilization.

A high voltage power transmission line is being built from Kenora to Ignace and I hit the pole line construction trail. It's bull work to start with, but it pays well and I need the money. I started 1971 with 10 grand in my pocket but all the bills have come in and I'm almost 10 grand in the hole now. Those people trusted me and it's time to knuckle down.

In two months' time I work my way up to Nodwell driver, and then the whole crew wobbles! Crap – one step ahead and half a step sideways!

There is a pipeline compressor station 15 miles west of town and there is an opening for an operator. I apply and get the job. It's a straight-up deal – 40 hours a week on a 7-day swing shift, leaving me lots of free time to fill.

The local propane distributor needs a delivery driver, and now I have two jobs. I'm running 90 to 95 hours a week between the two outfits, and I might as well be in the Arctic as far as my family is concerned. It's like Elmer and his standing dry poplar firewood. My wife and I meet occasionally at the back door. (She is also on shift work.)

Two years later I hand deliver the last cheque to my last creditor, the general store in Savant. Everyone is paid and no one can be happier than I. All of my creditors have been patient throughout, and not one of them asked for interest. They all appreciate my effort.

I quit the propane job, and now I have all this free time on my hands. How will I fill the empty hours? Back to the bush, of course. Go figure!

Table Scrap

During those two years of bill-paying, I got a little help from my banker. It was sort of an "Oops!" deal.

By 1973 I had been banking for four years at a major charted bank in Dryden. Early on my account had been very active – now not so much.

I got a letter from them – why not stop in and chat with the loans Manager? They would be glad to lend me up to $3000.

So of course, why not stop in? I did so.

The loan manager was fairly young and all business. I had aged a tad in the last three years and was out of business. We both laid our cards on the table and I hid nothing. I told him I was still five grand in the hole and needed some consolidation. He asked me for some security. I told him I had a '68 Impala with 150k on the clock, road rash creeping up the doors and two jobs. He said, "No dice."

169

I was a bit peeved now, and I asked him if he was in the habit of sending out letters to folks like myself just to jerk us around. I asked him if his middle name was "Lucy" – was he holding the football – did I look like Charley Brown? I said maybe I should talk to the head guy or write a letter to Toronto. He caved in, gave me the three grand and actually took the Impala for security.

It was this loan that allowed me to write that last check to Emerson Ennis and quit one job six months later with a small easily-handled balance owing at the bank.

An ex-banker friend explained the deal. Every year an audit is done on the loan guy's portfolio. He is supposed to have four or five percent deadbeats on the list. If not, the bank wants him to be more aggressive.

What a strange business model I'm thinking – built-in failure!

I must have really buggered up his career path when I repaid that loan.

It's tough to go cold turkey when fellow addicts won't leave me alone. My reputation is still blowing around the tops of the jackpines and people keep stopping by.

One evening I drop in at the bar for beer and shuffleboard and strike up a conversation with a gent I have never seen before. We swap lies for an hour and gravitate to my house for a nightcap. Before the night ends the truth comes out. The guy is a Falconbridge district geologist, and he has heard about me. It turns out that this whole evening has been an evaluation process, and they are looking for a geophysical foreman. He likes the cut of my jib and offers me the job. I am tempted, but respectfully decline the offer. This has to be the weirdest job interview of all times and I still chuckle when thinking about it.

Charlie Miles and his partner Mike look me up. I don't know them from Adam, but they are in the area and they've heard about me. Both these guys are in their mid-fifties and have been banging pieces off rocks since they were knee high to a grub hoe. We have them over for dinner a couple of times and spend the evenings swapping stories, and these guys have tales by the ton – here's two of them.

Sometimes You Just Get Lucky

Charlie and Mike had spent the summer working the shoreline of Hudson's Bay north of Rankin Inlet – no specific target, just checking out old and new showings of interest. They have an Eskimo guide with them and are using a lake freighter canoe. (Lake freighters are wider, deeper and stronger than their inland cousins. They are square-sterned and can handle ocean swells pretty good – but the boys still hug the coastline).

It's late August, and they are working their way back to Rankin. They are still three days out when they stop to check out a gossan on a little island they had passed when heading north three months ago. It's a beautiful, calm, warm sunny afternoon, and they have been chugging along in their shirtsleeves, watching the shoreline geology as they cruise past, pretty much enjoying life.

They reach the island, find a good spot to disembark and do a little prospecting. Charlie says they always tie off the canoe, but today they don't plan to stop long – the Eskimo sits on the shore holding the rope.

The gossan is pretty interesting, so they do some sampling. An hour later they return and what do they find? – the Eskimo sleeping in the warm sun with an imaginary rope in his hand, and their canoe bobbing in the water 20 feet away!

What to do? Mike says he will swim for it. He disrobes, steps into the water and locks up – Charlie and the Eskimo pull him out. They go plan 'B'. They take off their shirts and pants. They will tie their clothes together, put a rock at one end and cast for the boat. Before they can finish their improvised rope a breeze springs up and the canoe sets sail for the mainland 200 feet away. (Charlie says if plan B had been plan A, they probably would have snagged the canoe.)

The weather deteriorates and by nightfall a strong wind is blowing in, waves are crashing over the little island and it starts to rain. A cold wet night is in store, and they have no shelter. The Eskimo is better off, he is still in his sealskins. Charlie and Mike spend the night stomping around and swinging their arms, fighting off hypothermia.

The day dawns clear but cool, and the wind continues. They find a pool of fresh rainwater on the island, so at least they can hydrate. Their biggest problems are cold and hunger. To top things off they can see their canoe, containing their warm clothes and grub, bobbing in a little cove 200' away and sheltered from the waves in the lee of their little barren island. The Eskimo keeps his distance and avoids eye contact. He's heard the elders tell their tales of the old days. The bad guy gets eaten first.

No one trots around the arctic willy-nilly. Rankin has their itinerary and knows they are due in two days, but how soon will they react? It will probably take a day before they start to search, and can Charlie and his cohorts last three days? They sit and watch that dam canoe, two hundred feet and a lifetime away.

By golly! Later that afternoon a plane flies over on its way back to Rankin and it spots them! It is a Grumman Goose amphibian, but it can't land – the ocean swells have not yet subsided. The plane makes a couple of low level passes and waggles its wings. "Hang in there," says the Goose, "We'll come back for you."

It's another miserable night on the island, but the situation is not so hopeless now, although the next day is not much better. The wind is still blowing unabated, and it's cold, cold, cold!

By 2 pm the boys are about to give up when they spot smoke on the south horizon. Rankin has sent out a small coastal freighter to rescue them, and two hours later, Charlie, Mike, the Eskimo and their canoe are loaded aboard. Salvation!

When Charlie finishes his story my skin is all prickly. Truth will trump fiction every time!

Table Scrap: Getting Lucky Part II

It's another year and another season. Charlie's syndicate has picked up the rights on a molybdenite property near the old Black Donald mine in the Calabogie/Killaloe area west of Ottawa. Back in the day a shaft had been sunk to a depth of 150 feet and some drifting had been done. The boys want to check the ore grade in the lateral workings, and to do so it is necessary to dewater the 10'x10' square shaft.

They have a simple straightforward plan. They build a substantial raft small enough to fit in the shaft but large enough to hold a 3-horse water pump. As the water level goes down, the raft and pump follow, sections of discharge hose being added as necessary.

171

This will be a 24-hour a day job and Mike will run the night shift. They live in a tent nearby.

The routine is thus: the pump runs on gas and a tank lasts for about two hours. When the motor quits they wait for a certain amount of time to allow the carbon monoxide to dissipate before going down the shaft to refuel the motor. A steel ladder was rock-bolted to the shaft wall back in the day and it is still in good shape. This is their entry/exit route.

They have dewatered 100 feet of shaft now, and Charlie waits for an hour before heading down the ladder to refuel. He fills the tank, fires up the pump, heads up the ladder, and 30 feet up he realizes he has forgotten to check the oil in the crankcase. He climbs down, stops the motor, checks the oil and starts the pump again.

Half-way up the ladder Charlie knows he is in trouble! He should never had gone back to check the oil. He is getting woozy and has 50 feet to go to fresh air!

The last thing Charlie remembers is seeing blue sky 30 feet above him. Half an hour later he wakes up and he's lying on the grass and his feet are hanging over the edge of the shaft! He's retching, coughing, and not at all that healthy, but he is alive. Once again I feel all prickly - WOW!

They say that the only people who never make mistakes are those who do nothing. Charlie said he made many a mistake in his lifetime.

I wish I could recall more of Charlie's tales, and I wish I had kept in touch with the two of them. No matter – I sure enjoyed that short interlude in the seventies.

Sidebar: Charlie was from Whitefish Falls Ontario, a little community on Manitoulin Island not far south of Espanola.

At that time there was a series on CBC-TV called "Rainbow Country." I always thought the show was bogus. The central character, a 15-year-old boy, had an English accent. (Like they couldn't find a Canadian lad who spoke English.)

Anyway – when they ran the opening credits (an aerial view of Rainbow Country) you could see Charlie's house in a secluded bay. Pretty neat, hey?

I stake six claims on that conductor on the northwest corner of the Bridges Twp. grid that Noranda had deemed uninteresting. I put in 6 miles of picket line (working on my days off) borrow Kerr Addison's JEM and mag, and plot a nice little conductor with flanking mag association. Kerr Addison Jim is willing to take a crack at it, and pays me $600 on an option agreement. One 4-day long weekend I take Kerr Addison's Winkie drill in and with the help of a local guy I put down a 100-foot hole. I even spot the hole myself, but darn it all anyway, at bedrock I collar the casing in sulphides, and can't save the core. I grind away some stuff and the rest of the hole is relatively uninteresting, with a two-inch band of sulphides at depth. I want to pull back 50 feet and try again to get a second intersection, but Jim won't go for it. It's ok though, I had used their equipment, received 600 bucks on the option, and they even paid me to drill it. All it cost me was some time, a little bit of travel, and some sweat. For a change I am on the black side of the financial ledger.

I submit the property to a half dozen other outfits. I get answers from three of them, and they are uninterested.

(Want to know how to let a prospector down gently? See inserts.)

CANADIAN NICKEL COMPANY LIMITED

COPPER CLIFF
ONTARIO

June 12, 1973

Mr. R. Durnin
Box 91
Vermillion Bay, Ontario

Dear Bob:

We have reviewed the data supplied
by our geologist J. Hannila on your property
near Vermillion Bay, Ontario. The work done
on the property to date has not been encouraging and therefore we are not prepared to
enter into any exploration agreement at this
time.

Thank you for submitting the property for
our consideration. The data you loaned Jorma
is enclosed.

Yours sincerely,

J. E. Mullock
Regional Manager
Field Exploration

JEM:vs
Enclosures

RAM PETROLEUMS LIMITED

P.O. BOX 90 THE SIMPSON TOWER • 401 BAY STREET • TORONTO 103, ONTARIO • TEL. 863-1010

AREA CODE - 416

CABLE ADDRESS "RAMPET"

August 16, 1972

Robert R. Durnin, Esq.,
P.O. Box 91,
Vermillion Bay,
Ontario.

Dear Bob,

Re Claims - Bridges Township

My knowledge of geophysics is very limited but we had your information reviewed by a consultant. His conclusions were in brief:

(a)   that he did not consider drilling a check hole at depth a good risk.

(b)   he did not like either instrument that you used.

Consequently Ram would not be interested in following it up. Have you shown it to Kerr Addison? I enclose your plans and reports, etc.

Regards.

Sincerely,

C. C. Allen.

CCA:SPI

Encs.

174

TELEPHONE 269-7991   P.O. BOX 1448

SCURRY - RAINBOW OIL LIMITED
709 - 8TH AVENUE SOUTH WEST
CALGARY, ALBERTA   T2P 1H5

April 16, 1973

Mr. R. Durnin
Box 91
Vermilion Bay, Ontario

Dear Mr. Durnin:

Please find enclosed your report on the Cates-Tibor property in the Bridges Township of Ontario.

Scurry-Rainbow Oil Limited is not interested in this property at the present time.

Thank you for submitting your report to us for consideration, and we apologize for the long delay in returning the report to you.

Yours very truly,

SCURRY-RAINBOW OIL LIMITED

Lorraine Lee

for W. C. Cheesman
Manager of Exploration,
Mining Division

/lj

Encls.

KERR ADDISON MINES LIMITED

SUITE 1600 · 44 KING STREET WEST
TORONTO 1, ONTARIO
TELEPHONE 362-2608

May 24, 1972

Mr. R. Durnin,
VERMILION BAY, Ontario.

Dear Bob:

Enclosed please find a copy of the drill log for the hole drilled on your claims in Bridges Township by Kerr Addison, March 30, 1972. Also enclosed are the transfers for the claims, and copies of relevant sections and plans.

I am sorry that results did not warrant additional work, and that we have to advise that we are discontinuing the option agreement. It certainly was an interesting target, and well worth the test.

Best of luck, and I hope we will have an opportunity of working with you again in the not too far distant future.

Yours very truly,
KERR ADDISON MINES LIMITED

G. M. Hogg
Chief Geologist - Exploration

GMH:lfr
Encls.

cc: S. M. Sober

It's another long weekend and I'm off to Savant Lake again. Kerr Addison Jim wants ten claims staked and I agree to do so. The only problem is that the claim block is along the C.N.R. line five miles east of Savant and there are no roads or water access. I will have to walk in.

Savant Lake hasn't changed much. Ontario hydro has hit town, and there is no longer the hum of diesel generators. There is even talk of a new motor hotel in the future. I check into the old hotel, renew acquaintances and it's just like old home week – pretty nice.

I look up my old friend, C.N.R. Lloyd, and as we chat a solution to my claim access problem materializes. Lloyd is working the east line for a couple of days and if I can keep my mouth shut, he will pick me up at the edge of town and I can ride along on the gas car (a.k.a. jigger.) This is great! I have to walk home in the evening but I can handle that. In the bar that night I manage to control my tendency to yap and the next morning me and my staking axe are waiting on the main line just east of town.

File pic of a jigger similar to Lloyd's.

This is my first ride on a jigger, and believe me, this is no Buick Roadmaster. The dam thing is not much more than an outhouse on wheels. The cab is barely wide enough to hold Lloyd, me, and his helper on the bench seat, and we careen down the track, bobbing and weaving at 90 miles per hour (I'm sure) with nothing between me and certain disaster but a windshield. The gas car is not much more than four feet long, seems to be eight feet high, and I'm sure that if we hit a groundhog we will tumble end over end. Lloyd tells me that he might hit 40 mph downhill with a tailwind, but the blind corners come up pretty darn fast – and what if we meet a train? It beats walking, but not by much. I don't mind trudging back down the track in the afternoon.

Lloyd drops me off every morning and it takes me three days to do a two-day job. My work at the pipeline is mostly maintenance and gauges, and I'm out of shape. The first day my wristwatch flops around and my hands develop blisters. The second day I can hardly hold my axe, the muscles of my forearms are screaming, and my hands are bleeding. By day three calluses replace the blisters and my watchband is too small. My watch goes into my pocket. Oh to be young again!

With the staking done I leave Savant, but I will return. I have a legend to track down, but it will wait for next year.

Table Scrap

Lloyd tells me that there is little danger of meeting a train if one pays attention. Before heading out, Lloyd sets up a "block," meaning that he knows where the trains are and how far he can go before pulling off the track. All my life I have seen little wooden platforms at intervals along the railroad right-of-ways, and I find out that these are the spots that Lloyd and his ilk can use to park the gas cars. The unit has handles at the back and is not very heavy. Lloyd just grabs the handles, swings the rear wheels onto the platform, pulls it off the track and waits for the train to pass.

One day Lloyd cheats and rolls the dice. He has made good time and decides to run the block. He can make it to the next pull off, he thinks.

Well, I guess the train was making up time also, because Lloyd rounds a curve and not to far off a locomotive is coming, and somewhere in between is the little wooden platform. It's time to sink or swim!

Lloyd slides to a stop right on the money, jumps out, grabs the handles and yanks the jigger off the track. He falls back pulling the gas car over his chest, and just as it clears the rails the train whistles by! Lloyd never runs a block again.

Table Scrap

A sort of no-man's land exists along the C.P.R. line between Kenora and East Hawk Lake. A high voltage transmission line and natural gas pipeline follow alongside the rails, and the only access is a rip-rap trail along the pipeline. About halfway down the boondocks the pipeline has a rectifier ground bed and it's on the wrong side of the tracks, so to speak. The pipeline is doing some maintenance work on the ground bed, and the battery is dead on their backhoe. The boys are too lazy to carry in a fresh battery, so they decide to jump the tracks with the crew cab. The truck high-centres on the rails and a freight train comes around the corner. The men bail out, and with hard hats, tools and pieces of Dodge flying around, the poor engineer thinks he's wiped out everyone concerned. No damage other than mangled tin, but I'm sure the train crew aged a few years.

The legend has been floating around Savant since 1969, and I've been intrigued for years. Before I left Savant Lake after staking the ten claims last year, I had talked with Billy Read, who is reputed to be in the know. Billy is retired from the Dept of Highways now, but his wife still runs the post office, and I've known them since 1971. They are good honest folks.

Billy confirms the legend as almost, maybe, perhaps, true. It seems that a geologist (name withheld) working for a company (name withheld) had pulled an intersection of massive chalcopyrite near Fitchie Lake, about 25 miles north of Savant on Hwy 599. The geologist knows that his company is only interested in nickel, and is likely to pass on the chalco, and besides that, the guy is pissed at his boss. He is planning to pull the pin anyway, so when they yank the drill out, as legend has it, the core box containing the intersection accidentally on purpose falls off the muskeg tractor, spilling the drill core into a deep creek never to be seen again. I'm going to check this out.

I visit with Kerr Addison Jim in his Kenora office quite often, and during that winter I share the legend. Jim is equally intrigued, and we reach an agreement. If I can pin down the area and stake it, Kerr Addison will take an option on the claims. I touch base with Billy again and he narrows the location down pretty well. I tell him I will see him next summer.

Summer arrives and on a long weekend I tie my fibreglass canoe on the Impala roof racks, toss in my staking axe and bush boots, and head to Savant. Bob and his wife have sold the hotel and are now enjoying a second retirement in a cabin on Sturgeon Lake. I don't know the new owners, but there is a spiffy new motor hotel just down the street. I check into the Four Winds, and it's clean and modern, and the owners are nice folks. I like it fine, but go figure – I kind of miss the old place with its sashed windows and the huge clawfoot-tubbed bathroom down the hall. I guess I'm already a curmudgeon.

Sidebar: In 1968 Bob and his wife were empty nesters with a plan. Bob would take an early retirement package from his position as a store manager at a national retail appliance company, and they would buy a nice, quiet hotel in a nice, quiet little town, and relax and enjoy life. Talk about bad luck! First Mattagami Lake hit zinc on the south end of Sturgeon – ditto Umex at Pickle, followed by road construction on 599, an expansion in the logging industry, tourism and what-not! Bob and his wife were busier than bees, and by 1975, they were tuckered out and peopled out. They sold the hotel and built a four season, secluded cottage. Some people have no luck at all!

To reach my staking target I have to travel seven miles east on Fitchie Lake, portage a half mile into Solitude Lake, and thence by canoe, two miles south to the end of Solitude. Solitude I can handle, but I'm far from a world-class paddler. I need a boat to cross Fitchie, so I go to see the semi-bandit (the same guy who rented me the little house in 1971.)

(Semi-Bandit could have been cast as Haney on Green Acres. You recall Haney? The crafty "If I don't have it I'll find it for you" guy? Well, this guy is Haney in the flesh.)

Yes, he has a boat, and it's just what I need. It's a 16 foot aluminium fishing boat with a 25 horse Johnson, and it's tied up at the government dock on Fitchie right now. I can step right in and motor away, so I pay him for a tank of mixed gas and two days' rental, throw the Cruise-a-day into the Impala and head north.

There is a provincial park on Fitchie Lake and in one corner a road construction crew has set up a camp – they are upgrading a 20 mile section of 599. I pull into the parking lot near the dock, unload my canoe, carry it to the dock and find my boat. Kind of strange though; there is a gas tank and a couple of fishing rods in it! I check it out – yup, there is Semi B's name on the boat, and the number is correct – it's my boat all right. I switch tanks, lay the rods on the dock, tie the canoe lead rope to the transom, and before I can step into the boat someone hollers at me. "Where in the hell are you going with my boat?"

I turn around and see a guy with a woman and two kids behind him, striding down the dock, and he is eye-rate with a capital Eye! I ask him what he is talking about, and he says he has the boat rented for the summer, and I can get the hell out of here. Semi B, the shifty snake, is double-dipping with his 16 footer.

I size the guy up and I'm thinking fast, and usually I'm not the swiftest pony in the corral. I've got 12 inches and 100 pounds on this little guy, but I know a thing or two. You don't run a construction crew by being touchy-feely, and this guy looks like he can handle himself, if you can catch my drift. I need backup, so I reach into the boat and grab my staking axe. We are now facing off in the North Atlantic. I have the missile in my hands and he's not sure if I'll push the red button. He steps back a couple of feet and the negotiations start.

He has his family at the park for a few weeks and he has promised to take the kids fishing today. I have driven almost 300 miles to stake a couple of claims, and I don't have time to rustle up another boat. He wants the boat, but I have the axe.

The lit fuse sputters out. He heads back to dry land muttering about Haney. I climb into my commandeered pirate prize ship, and motor off. About 100 feet out I glance back. Two little kids are standing at the end of the dock watching their fishing boat disappear. I don't look back again.

I carry the canoe into Solitude and it's uphill all the way. The bush changes character as I walk, and even before I reach the lake, I know I'm in for a treat. At the top of the portage I drop the canoe into a beautiful, crystal clear, spring-fed body of water, a half mile wide by two miles long. I paddle down to the south end and every inch of the way I can see lake bottom, no matter how deep. Solitude is surrounded by jackpine covered sand flats, is absolutely pristine, and very well named. It's eerily quiet here, and the outside world seems far away.

The bush is like parkland and it doesn't take long to find three diamond drill holes. The casings have been left sticking a foot or so above the ground, and with no underbrush to speak of the holes are easy to spot. I stake the perimeter of my six claims, but it's getting late. I will return to finish the job tomorrow.

I leave the canoe on Solitude, tie up the boat at the government dock and take the Cruise-a-day with me; I'm not 100% sure I will have a boat tomorrow. I may have to bum a ride to the portage and paddle out later. Right now, I'm going to pay a visit to Haney.

Semi-B is unconcerned. Haney has no scruples whatsoever, and the more pissed I get the more he laughs. You couldn't make this guy blush with a can of red spray paint. He claims he has stroked the foreman. I hope he's telling the truth for once in his life.

The next morning I meet the foreman on the road into the park. He passes without looking at me and I feel a little bad about it. Not too long ago I would have brought in a 12 pack and sat down with the guy, but I have no time right now for such shenanigans.

Whew! The boat is still at the dock. I bang off the rest of the staking by noon and I'm back home in Vermilion Bay in time for supper that evening.

On my next work break, I drive to Sioux Lookout, record the claims, and take the map to Kerr Addison Jim in Kenora. He gives me a standard option on the six claims and writes me a cheque for $600. I send Savant Lake Billy 100 bucks and have 300 bucks profit for three days of enjoyable work. Better than a sharp stick in eye, I'm thinking.

Well, wouldn't you know it? A little more than a month later, and I'm on my way back to Savant.

When it comes to loose lips sinking ships, Kerr Addison Jim could send the whole U.S. Navy to the bottom. At his daily lunch with the Kenora Fraternity of Deep Pockets he has been over-enthusiastic about how he lucked in on a hot copper prospect. One of the guys is going to send a staking crew in to the Fitchie Lake area. It's time for Jim to cover his butt.

He needs a 50 claim donut staked around the original six, and he gives me the contract at 50 dollars a claim and Kerr Addison will pay the recording. Jim even throws in a 3% N.S.R deal on the 50 claims. Sweet!

I take a week's holiday and head back up 599. I hire Rosaire, my old tried and true staker, from Umex days to help me. Rosaire lost all his fingers on his left hand in an industrial accident a couple of years ago but he can swing an axe with one hand better than most men with two. Rosaire is a helluva man.

I have to deal with Haney again – he is the only game in town, but the highway construction is done and the boat situation is hassle-free. We finish the job in no time flat and Jim pays me 2500 bucks.

How ironic! I left the bush because I couldn't make a decent living at it. Now, every time I step off the pavement I go home with money in my pocket - go figure!

Jim gives her a good shot. He hires a consulting geologist to do an independent report on the area and the consultant is equally enthused. He feels the whole area arcing east and north along the adjoining Neverfreeze Lake is comparable to Flin Flon. Mr. Kenora Deep Pockets stakes more than 200 claims, tied on to us and along the Neverfreeze trend. Holy cow! Billy and I started our own mini-rush!

Deep Pockets' staking contractor was none other than my old friend Don MacEachern. As an added bonus (no doubt due to my friendship with Kerr Addison Jim) I get a 1%NSR on those 200 claims. I now have an interest in over 250 claims, but I hold off ordering my El Dorado convertible. Any percentage of zero is still zero.

Deep Pockets puts a drill in, Jim also puts down five holes, and no one comes up with diddly-squat. They don't make legends like they used to, I guess. Things fizzle out and I say goodbye to Savant Lake, and I've never been there since.

In 1979 my wife pulls out. I really can't blame her. She spent 13 years trying to domesticate me, and it can't be done. The problem is, I now have two pre-teen girls to raise, and I'm still a kid myself. I hang in for six months, but the memories are too much. I rent out the house we bought four years ago, divest myself of most of our worldly possessions, throw the girls into my old station wagon, and head for the Peace River oil patch. It's time to turn the page.

Chapter XI   1980 - 1988

After a few false starts and hiccups I land a job with another pipeline and move to Winnipeg with my two daughters. Life calms down a bit but I'm still a work in progress. My parenting skills are limited – that was always left up to my wife. My girls are reaching puberty and I am awash in a sea of female hormones. My older sister is a rock to lean on but she lives 40 miles away. We struggle a bit, but life is not too bad. There's always food on the table.

Food! That's the stuff you find at the supermarket, right? Maybe you shop for your groceries, but since 1960 my grub has often literally materialized out of thin air, delivered by a Beaver or Cessna 180. It may be shopped for at times, but I have always delegated that responsibility. Sometimes it had arrived in the back seat of my wife's car. I helped carry it in – ergo, I have shopped.

So here I am on the wrong side of forty and for the first time in my life I am pushing a grocery cart! No kidding, never before have I done this – and crowds? I can't handle them! For 20 years I have been a loner preferring my own company and comfortable in a three-man tent. On the sidewalk I no longer count paces; but it's a good thing my girls are on the ball. I am still dysfunctional in civilization!

Here's how I shop at the Superstore. I start at one end and work my way south on the grid. The trouble is, the store is too darn big, and the cart is too darn small. If I start at the fruit and produce end, I never make it past the Alphagetti. No matter - this payday we eat healthy, lots of roughage and corn flakes. Next payday I start at the bakery end and pass the meat counter before I power out. No problem – now we will be carnivores for two weeks. There is still a couple of dried up apples and wrinkled cucumbers at home, and don't ever start me in the houseware section. I am sure to reach the checkout looking like the Fuller Brush guy, and who needs a list? Not Super Shopper me, that's for sure. Even now, if my wife sends me for three items, I'll come back with six, only two of which she needs. (Two out of three ain't bad.) One thing I've noticed though – she has started to pin notes to my shirt pocket before she sends me off. She might have been born at night, but she wasn't born last night.

So I'm living in a city of 500,000 + and the bush is only a distant memory right? Wrong! Kerr Addison Jim is in town. He is just Jim these days. Noranda shut down the Kerr Addison exploration arm last year and Jim is on his own. He is now doing some consulting and is struggling to jump through the regulatory hoops as he forms his own junior resource company. I renew an old friendship and it's really great. I visit him often and even run into Carl Huston who was last seen driving back to his Winkie drill with his upright Econoline van in 1971. Cool, I will visit Jim often over the next few years.

In 1981 my girls are spending most of the summer with their mother, and Jim wants me to put in a grid and do a mag survey on a 20 claim block near Atikokan, Ontario. I throw my axe and packsack into my old wagon and head for familiar territory – a working vacation.

I rent a cabin and a boat at an old fishing camp on Perch Lake, fifteen miles west of Atikokan. The claim block is six miles up the Seine River and the southwest corner is not far into the bush at the foot of Boyce Rapids.

Perch Lake is part of the Seine River system which runs from Atikokan 40 miles or so west, where it empties into Rainy Lake at Seine Bay near the old gold camp of Mine Centre. The river widens into at least three lakes along its travels and is a beautiful stretch of water. I urge you to visit it someday.

Boyce Rapids is a mile-long stretch of fast water, but is navigable in a 16' boat with a 9.9 motor if you are careful. By running the rapids I can access the northeast corner of the grid. I go to work.

Sidebar: The Ontario Department of Lands and Forests at some point in the past has limited certain waterways and small lake systems to under ten horse power, (not a bad idea really) and to meet these standards, every outboard manufacturer markets a 9.9 horse motor. I think they just manufacture a decal and slap it on any popular fishing motor. It could be 7 hp or 15, but if the motor cover says 9.9 then 9.9 it is. One day I studied one of those decals with my prospector's magnifying loupe and in very small, fine print under the 9.9 were two little words "Bite Me."

On my second day at camp I meet Don Huston, and Canada shrinks once again. His dad is Econoline-Van Carl and Don is also doing some geophysical work on an adjoining claim block. He calls his dad and tells him I am in camp, and his dad says he'd better take care of me. He tells Don about the rescue job in Savant 10 year before. Don is a great guy, and we tip more than one jar over the next two weeks.

Jim comes down from Winnipeg. He is going to spend a day banging pieces of rock off some outcroppings, and we head up river the next morning. The weather doesn't look too promising, so we have a plan. We will hit the bush at the west end of Boyce on the south west corner of the grid and if it starts to rain before noon, we will meet here. If the rain holds off until later, I will pick up Jim at the head of the rapids. Wouldn't you know it? It starts to rain at noon on the button.

I trudge back to the boat to wait for Jim, but which way will he go? It's raining harder now. The temperature is dropping and my jacket is not waterproof. I'm getting cold and wet and still no Jim. Maybe the poor guy is shivering at the top end of Boyce, so after 20 minutes I carefully chug up the rapids to check in at the southeast grid corner. I call, but to no avail. Maybe he came out below? I wait 15 minutes and motor slowly back down to the other corner. This goes on for two hours, and I've made at least eight round trips. I've got Boyce down pat by now, and the little 9.9 is kicking up a rooster tail as I speed wide open between the rocks. Finally, Jim emerges at the top of the rapids and he is dry as a bone. That crafty old Scot had a plastic rain suit in his packsack. I'm pretty cranky, and I tell him he's cheating. Jim just laughs like crazy.

We head back to camp, and I'm so cold I think I may die of exposure! We beach the boat and I hit a hot shower, and in ten minutes I'm fully recovered and in dry clothes. Civilization sure has its merits.

Two weeks later I am heading back to Winnipeg and I stop in Mine Centre to visit Harry Bell, the mining recorder in Kenora back in the seventies. Harry is now retired and he and his wife live in a spiffy self-built log house on Bad Vermillion Lake. I've known Harry since I was a callow youth and I sure enjoy the visit. Harry is the unofficial custodian of all things historical in this old gold camp, and I wish I had more time. I continue on to Winnipeg feeling all warm and fuzzy inside.

Sidebar: I first met Harry in 1955. He was a clerk in the local Co-Op feed store in Fort Frances then, and I was an A-hole sixteen-year-old with <u>one</u> proud possession – my new driver's licence. My only claim to fame was the farm pickup, and I drove the wheels off of that sucker. Thus, most of the adults in my world treated me with disdain – but not Harry.

Occasionally I stopped at the Co-Op to pick up a bag or two of supplement for our dairy herd, and would deal with him. He was always friendly and interested in what I was up to.

At first I was perplexed – I was usually up to no good. But Harry was non-judgemental, and every time I left the feed store I left feeling that maybe – just maybe – I was a little bit worthwhile after all.

Harry moved on to be assistant/trainee as the local mining recorder – Roscoe Richardson would soon retire.

When I hit the bush, I would often run into Harry. He moved to Sioux Lookout and on to Kenora and I would always receive an old-time warm reception when I stopped in to record claims.

If things were going poorly I would sometimes visit the Bells for an evening coffee. I never complained or whined to Harry – it was always just a pleasant evening of conversation. The thing was – when I left the house I always felt tomorrow would be a better day – and thanks to Harry, it usually was. I loved that man.

Back in Winnipeg I turn in the maps to Jim. But because of bad weather and dirty bush, the job is only half done. Jim understands – the old coot might pinch a penny until it hollers, but he sure has always been good to me.

My fortunes take a sharp upswing. Probably because I've quit looking, I meet a new lady, and she is a dandy. I sure like her, and to my amazement, she likes me! I figure she's one of those people who have no fear of the unknown – she'll learn.

The relationship progresses swiftly. At our age there is no time to be wishy-washy – it's poop or get off the pot time. Soon we are shopping for six, four of whom are ravenous teenagers. Now I push the drag cart in the in the grocery wagon train, and things are back to normal – food just materializes in the cart. I buy a pair of mirrored sunglasses so I can ogle pretty young housewives as they squeeze their melons. Life is so good!

The following summer the kids are at our respective ex's, and we are heading to Atikokan. No bush foolishness this time, though. We are going to spend two weeks at Perch Lake doing what normal folks do. We are going to relax – at least I am.

A few years before, a new highway had been built joining Dryden to Hwy 11 east of Fort Frances. I've never driven it, so after two coffee stops with friends in Vermilion Bay and Dryden, we take the Trans Canada east and head south. We pick up a cold six pack in Dryden and drive leisurely, sipping on a can and enjoying the scenery. At the Hwy 11 junction I turn left to Mine Centre. She doesn't know it yet, but we are going to visit Harry Bell.

One thing she does know – I'm in my mid-forties now, and already have a cranky prostate. One beer equals one rest stop, so as I pass Mine Centre and say I'll find a side road to use and she's cool with that. I turn onto the road leading to Harry's driveway, but I can't find the darn sign, and I drive past. I turn around and head slowly back, and as we round a little bend she spots the sign nailed to a leafy poplar.

It couldn't have been scripted better by a Hollywood screenwriter. She hits that bait like a starving muskie, and cries, "The Bells, The Bells, let's go visit the Bells!"

"Fine by me," I say, and turn in towards Harry's cabin, and the road is kind of turkey-trackish. Harry has never owned a car, and couldn't give a rat's behind about his road.

We cross a dry creek bed with the back bumper dragging and my wife is getting agitated. "What in the hell are you doing?" She cries, (Her vocabulary has expanded somewhat since she met me.) "You can't just drive into some ones yard!"

"It was your idea, not mine," I reply, and pull up at Harry's cabin, and he's already heading to the car with a big smile on his face.

"You sonova bitch!" she says, "You know him don't you? Is there any place in this blankety-blank country where you don't know someone?"

"Not many," I say, and I laugh like crazy.

We have a dandy visit and head on to Perch Lake. My wife loves to fish – me not so much. I'd rather cruise the lake checking shoreline outcrops and admiring bush I don't have to walk in. So we compromise – we run at half throttle. I'm itching to see what's around the next corner – she's watching a minnow water ski behind the boat. The only fish she catches is a scrawny old Jack, too old and skinny to duck the hook, I figure. We enjoy our vacation immensely.

In Winnipeg I continue to visit Jim in his downtown office, and in 1985 both our lives take a left turn. My problem is a new terminal manager. Jim has problems with his board of directors. I'll deal with my problem first.

Our good old manager has retired, and he has been replaced by a hired gun whose mandate is to cut the operating budget, he has the concept of "figures don't lie but liars can figure," down pat.

His solution is simple and brutal. He will terminate four of us, thus cutting our $20/hr. wage from the operating budget. He will replace us with contract workers using a Saskatchewan pipeline construction firm. The construction company pays the guys $20 per hour and bills the pipeline at $35 per. He hasn't saved a cent but he has shifted some operating costs to the construction budget, which is naturally more flexible. Who knows what evil lurks the heart within… (cue maniacal laughter and symbolic wringing of hands.)

I don't have a hope in hell. Besides being the low man on the totem pole I don't like the guy, and I've never been good at hiding my feelings. The end is near and I know it.

One morning I go to work and the gunslinger meets me at the door with a self-satisfied smirk on his kisser. "Clean out your locker," he says. "You're done." I laugh in his face and ask him if he really thinks I'm that stupid. I tell him I sterilized my locker three days ago and he can have it surgically implanted you know where. I didn't even bring a lunch today, so I have no bucket to kick down the road ahead of me. Darn it all anyway, I liked that job.

I don't give up easily though. I phone the labour board to see if I have any recourse through that fine old government agency. Am I female? No. Am I a visible minority? No. Am I physically or mentally deficient? Well – I guess not, although I'm not too sure about the mental part. I tell the disinterested lady I'm just a middle-aged white guy who has been screwed out of a job. "Sorry – can't help you." Click.

So I stop in to see Jim. Maybe I can catch some staking or some such work, but before I broach the subject I learn that Jim has problems that make mine look insignificant by comparison. His venture company underwriting had gone through the hoops a couple of years before, and with Jim's wealth of knowledge and his reputation as a good honest geologist, things started out in high gear, but Jim is just too honest for his board of directors. They want to manipulate the stock by over-rating some of Jim's legitimate prospects to the detriment of unsuspecting shareholders.

Jim won't play ball, and his nice little venture company, started with such high hopes and such a bright future, has dwindled away. The stock is at rock bottom, and I mean rock bottom in big capital letters. It's still on the board, but it's at 1 cent asked, 0 cents bid. Jim is flat broke and is dejected – very dejected!

I ask Jim how this could happen and he tells me a tale that makes my blood run cold. When the directors found out that Jim wouldn't play their game they sat down and wrote cheques to each other until the treasury was empty! Jim is now putting food on the table by doing some consulting. His company is defunct.

I go home feeling bad for both of us.

(Somebody has to help me out here. When a company goes belly up it is defunct, right? So a healthy outfit must be funct, yet I've never heard of a funct company. No wonder English is such a hard language to master.)

My wife and I are not unprepared for the future. When we saw the storm clouds gathering we secured a mortgage and bought an 80-acre hobby farm 40 miles north-east of Winnipeg. Our kids are semi-independent now. The two oldest are on their own and the two youngest go to live with their other parents. It's not perfect for the two youngest but will have to do for now. The thing is – last year I grossed $45k and now I'm on Unemployment Insurance – a drop of $35k per year. I am negotiating with the pipeline head office in Calgary. I hope I can work at the terminal loading tank cars in busy winter months and hobby farm in the summers. It's a long shot and for the present my wife gets a job in Selkirk.

We buy a strong rear-tine garden tiller and plant a big garden. This will prove to be a good move. We also buy an excellent 1955 Massey-Harris Super 44 diesel with a front-end loader – also a good move. The Massey will clean out an old feedlot and remove snowbanks in the winter-time.

Sidebar: "Super" meant it had a 5-speed transmission instead of the standard issue 4-speed. I guess it didn't take much to be "Super" in 1955.

The Chicken Chronicles

We had a fairly large chicken house. The eastern portion had once been an egg-laying emporium but was now used for storage – lumber and valuables (junk.) The southwest quadrant had housed a few chickens five or six years ago and the northwest room was a dark, windowless mystery (which will be solved later on.)

For a few years now we had been sharing in the family chicken cooperative. My wife's father (they live forty miles south of us) had been custodian of our fall freezer herd. Every year we would all share in the purchase, feeding and slaughter of 200 cornish giants. Dad was getting on, and this year my wife and I would raise the chickens.

I cleaned the southwest quadrant and washed walls and floors with a water/lye solution until it was squeaky-clean. I then put down a layer of sand following with a layer of peat and topping it off with a layer of fresh wood shavings – I was going to do this right.

We put in 200 cornish giant chicks and watched them grow.

And they did right fine, soon becoming very active. So I built a chicken yard in the shade of some Manitoba Maples and put out some feed troughs.

The chicks, happy to be in-and-out, continued to do well for a while and then seemed to hit a roadblock. Then we started to lose some – one dead bird one morning followed by one or two every day.'

This was odd, so I took a carcass to a vet in Selkirk for an autopsy. They came up with a few ten-dollar words and recommended an expensive antibiotic. We were not too enthused about using drugs so I took two more carcasses to another vet. At $35 a shot this was getting expensive.

This tie the answer came back – cholera! How could this happen? I told the vet how careful I had been and he asked me if they had free-run. Yes they had, but the chicken yard had not been used for umpteen years. He then asked me if we had magpies around – yes we did.

Ahah – magpies carry cholera – it doesn't affect the magpies but it will kill young chickens. A simple antibiotic will cure and protect the herd but when we butcher them in September we will see pockets of pus on the livers. We can toss the livers but it may alter our perception of the meat – can we handle that?

All the ladies were consulted and the decision was made. The herd must be put down.

All my life I have, at one time or another, slaughtered animals for meat, but let me tell you – killing 175 partly-grown Cornish Giants was an emotional test. I felt I was being disrespectful of their spot on the food ladder.

The least I could do was give them a decent burial plot, so I dug a big hole at the north end of the alfalfa field. Then I wrung necks and threw the carcasses into the loader bucket – three trips to the cemetery. Then I backfilled and tramped the dirt with the heavy tractor. That evening I had a hard time falling asleep and I did not sleep well.

A week later I went back to check things out and found the grave had been robbed! The doggone coyotes had dug up the birds and not a feather was left behind. I went back with the Massey, levelled the ground and made a pass or two with the disker. Somehow I felt better now – the chicks had become part of the food chain after all.

Makin' Bacon

The wife and I were also taking over the annual pork smoking ritual and like the Cornish Giant fiasco it would run into a snag – but with a happier ending.

I had helped dad-in-law smoke bacon and ham for a couple of years. The smoker was an old steel-lined refrigerator with the compressor removed, a short stovepipe at the top and one of the original wire shelves to hang the meat from. It was a dandy smoker. There was room to hang two slabs of side pork (cut in half equalling four slabs of bacon) and two ham-hocks. Smoke from a slow fire at the base would percolate up past the meat and 24 hours of slow smoking did the job perfectly. Like farm-killed free-range chickens, home-smoked bacon and hams are far superior to store-bought.

Dad had taught me well. (?) The fire must be watched carefully, otherwise the dripping fat would burn and the fire <u>must</u> be kept under control. At nightfall sticks would be removed and the coal bed would slowly produce some smoke and heat until sunup. I had the procedure down pat – sort of.

So we brought the smoker to our hobby farm and made preparations. Our neighbours, Joe and Mary, had a little butcher shop in one of their outbuildings. We bought a whole pig and had chops, roasts and our side pork and front shoulders for ham.

We had a ten-gallon crock – the real deal. We put the side pork and shoulders down in a proper brine for ten days, and I collected smoke wood. Dad always used a mixture of old punky poplar, a piece or two of an old oak fencepost and sticks of green choke-cherry. I will add an extra flavour. Last winter the doggone mice had ringed (eaten the bark) of one of our apple trees and I had saved the dead wood.

Sidebar: The mice had killed one of our four apple trees. I found the cure in a farm paper and had I known of it a year earlier I could have fooled those rascally rodents.

It's as easy as walking in circles. After the first snowfall and thereafter until about Christmas, walk circles around the base of the tree. Mice will not burrow through packed snow.

We took the pork out of the brine and started the smoke process Saturday morning. All day I baby-sat that fire, keeping flare-ups under control and making sure that there was plenty of smoke and just the right amount of heat. That night the fire slept and so did we.

Sunday would be the last smoke day. Monday morning the cured meat would be hung in our cold room and on Monday evening we were to deliver one ham and two slabs of bacon to my wife's parents. Dad, we knew, was eagerly awaiting our first smoked pork crop.

Sunday also went well – up to a point. By mid-afternoon only four or five hours of smoke time remained. Mission accomplished (almost.)

Lloyd and Shirley (Savant Lake) who now lived east of Winnipeg were coming for supper, which they did, arriving at 3 pm, and that doggone Lloyd brought a bottle of rye.

So while the women-folk put dinner together Lloyd and I sat in the sun, sipped and talked, tended the fire and sipped, and sipped.

Supper-time was announced and two hungry lads needed no coaxing – into the house, the fire forgotten.

A half-hour or so later we pushed back our chairs, patted our tummies and Shirley, who sat facing the patio doors, said, "Is there a railroad track around here? Because it looks like there's a steam engine passing by."

Whoah! There was a funnel of black smoke drifting across our driveway! Out the door we dashed, yanked open the fridge and poured water on the fire and six hunks of burning pork. Just in time, too – a little charred on the outside with some trimming necessary.

Now I figured the bacon and hams were done so I closed the fridge and Lloyd and I went back to the house to finish off the bottle and the evening.

And two hours later Shirley hit the smoke alarm again! This time I used the garden hose, but it was too little too late.

The next morning I took what was left of one pitiful ham and sliced it open. Right in the centre was a perfectly cured piece, found, and just the size of a tennis ball.

I moved swiftly into damage control mode. A call to the abattoir at Beausejour found a freshly-killed pig. Twenty-four hours later we had two more hams and four more bacon slabs in brine. Seven days later they were hanging in the smoker and I baby-sat that sucker for 48 hours.

It tasted pretty darned good. We took Mom and Dad's share to them and Dad wondered why it had taken so long. We told him we kept the meat in brine for over three weeks to make sure it was special. Dad never said a word but I knew that he knew.

And we were never asked to make bacon again.

Smoke

Joe and Mary were dandy neighbours. They were an older couple in their mid-60's and had lived on their place ever since they were married – more than 50 years.

We hit it off immediately and became very good friends. It was out habit to visit on a Sunday morning, chatting and playing cards. They had 50 years of stories to tell and stories go better with Coke and 5-STAR. Sipping, talking and playing cards would eat up a couple of pleasant hours and then Mary would lay out a brunch. Sometimes it would be at our kitchen table and the 5-STAR, Coke and brunch would be on us. (I should mention that 26 ounces would last for more than one visit.)

One Sunday morning at Joe and Mary's our relationship took a bit of a turn. They knew us better now, knew we could be trusted and their 5-STAR came out as per usual. But wait – this rye had a deeper hue and tasted a bit different. In fact it tasted darn good and my wife and I raised our eyebrows.

Joe and Mary laughed. We were drinking "Smoke" they said, and we soon found out that they rolled their own, you might say.

Now the Sunday brunch stories got very interesting and although I don't remember all of them I will do my best, starting with some history.

Joe and Mary were wed on November 23, 1937. The wedding party came from the church to the farm by horses and buggies. It was a beautiful 80-degree day. (Global warming, anyone?)

Joe was taking over his father's place. His dad had died young so Joe and Mary built a granny cottage by the driveway and moved into the big (?) house. Granny would live in her cottage until she passed on. (Folks took care of their own in those days.|

It was subsistence farming on 80 acres. They had a chicken house, sold some eggs, kept a pig or two and milked a small herd, selling the cream. Joe worked out when he could. In the wintertime he often cut pine on the sand hills three miles east, taking lumber in lieu of wages. Thus he was able to build a neat barn and other outbuildings. He would also cut tamarack firewood and they would take a wagon-load of cut, split and cured tamarack to Selkirk on Saturday mornings. Mary and the kids (three daughters at first, followed by a son) used the cash to buy staples and occasionally a bolt of nice cloth for little girls' dresses. It was a simple, hard-working existence – no frills.

Sidebar: A truck from Selkirk picked up the cream once a week each time bringing last week's cream check – $3 to $3.50.

It was fly time and when the driver came to the door he saw that flies were a problem. "You need a screen door," he said.

Mary told him that screen doors cost three dollars – they didn't have three dollars to spare.

The following week the driver brought a new screen door on the truck. He told Mary that twenty-five cents would be deducted from the weekly check 'til the door was paid for. She could have kissed him. Pretty cool, eh?

Sidebar: When Joe and Mary were starting their family they had no radio. A lot of folks in the neighbourhood had no radio – radios cost forty dollars or more.

The furniture store in Selkirk had a salesman on the road. He visited Joe and Mary – they could buy a radio on time for five dollars a month. Joe and Mary were skeptical.

He was a super salesman. He brought the radio in and hooked up the batteries. He said they could try it out for a month at no charge and he left. Mary said it was wonderful. They listened to news, music and weather reports – but they knew they couldn't afford it.

The salesman returned a month later. He was told the radio had to go back and he was not pleased. He said they owed him five bucks for the month of listening pleasure – a classic bait-and-switch. They gave him the five, their total cash on hand.

Joe told us he watched the guy leave and he got mad – he needed that five dollars. He chased the car down their long driveway, reached into the open car window, grabbed super salesman by the throat – and he coughed up the five bucks.

So now it's 1945 and they decide to get into some on/off farm income – it's :smoke" time.

They partnered up with Mary's sister and brother-in-law who lived next door (where we now lived) and the mystery of the dark windowless corner of our chicken house is solved. It was the "smoke" room.

Joe and Mary's 80 acres ran east-west, one quarter mile wide on their side road and a half-mile down to Stan and Joyce's (our) place. S&J's 80 ran vertically, one quarter mile wide on their side road. J&M's north fence line Hit S&J's at the edge of the thick bush north of S&J's chicken house – got that?

The mash was laid down to ferment until cooking time. When cooking, the batch had to be closely monitored – they worked shifts, always one of the four doctors in attendance.

A visitor might drop in and let's say Stan was today's cook. Joyce would say "My husband is working in the chicken house." And of course, the visitors could not be allowed in there. "Have a coffee, Stan will be here shortly."

Then she would go to the phone. It was a party line hand-crank deal. A pre-arranged signal alerted Joe and/or Mary. One of them would boogie down the north fence line and through the bush to the smoke room. Stan would join the visitors at the house. Just cleaning the chicken house – innocent as all get-out.

They shared the product and each did their own marketing. Joe bought a '40 Willys four door sedan and expanded the laying flock. A route was established to Winnipeg. Each Saturday the back seat would be filled with cases of eggs. The trunk also held eggs but hidden behind them were gallon jugs of home brew. Back seat customers had eggs for breakfast – trunk customers had eggs and smoke.

Sidebar: I used to do the odd favour for Joe and Mary with my pickup. One day I asked him why he never drove a half-ton. "You can't close the trunk lid on a truck," he replied.

The wife and I were being tutored on Sunday mornings although it took a while to catch on. They told us how they gave the white lightning its unique color and taste. Smoke at the factory door must be diluted to be drinkable. They cut it with a light water and brown sugar syrup, cooked on the kitchen stove. Smoke served at home was never poured from a jug. It always appeared in a 5-Star bottle saved for that purpose and we donated our share of empties. If they were raided (never happened) the bottle looked legit. If the product was tasted the cops would know this was not your usual bland store-bought stuff. No problem – Joe and Mary would tell them that they were in the habit of adding coca-cola mix to the bottle. What a crafty pair.

They were raided once, actually. When their son got married the dance/reception was held in Stan and Joyce's (now our) machine shed. People came from far and wide and there was quite a crowd. Of course there was no bar – just pop and water available (wink wink.)

But the Mounties knew about the dance and the Mounties knew there was no bar licence and Mounties aren't stupid.

Now here's the deal. Behind the machine shed was a seed drill. In the drill box was a jug and a glass – one jug, one glass. The "bartender" was at the drill. When one wanted a drink one got one glassful after which the glass was rinsed with water. The jug, when empty, would be replaced by another stashed in the trees.

It was a good system. If someone appeared in the lineup too often and/or was getting a bit rowdy, they would be quietly led to the parked cars to have a nap in a back seat. The party would last until after sunup anyway.

The neighbour ladies were preparing the midnight lunch in the house. One lady came out to see Joe. "I just had a strange phone call," she said, "A man said, "Company's coming," and hung up."

And just like magic the jug disappeared from the seed drill.

Company never did come into the yard. They cruised the place two or three times and went on their way. It was just a heads up – and hocus pocus – the jug reappeared.

So we sat at Joe and Mary's and soaked up the stories along with the odd smoke. We'd known them a while now and had the nerve to ask them if they were "still" in business. They said they cooked the occasional batch to share with family and friends. We had already figured that out. The smoke co-op had ended ten years before and we knew they didn't have a gallon jug warehouse.

They had a small still buried in their bush lot. From time to time it would be dug up, set up in the now empty barn and a gallon or ten would be run off. The still would then be reburied.

And then – they looked at each other, nodded at each other and said, "We have two stills buried out there. We want to give one to you folks and we will teach you how to run it."

Well, Yankee Doodle! I almost fell out of my chair. We discussed it for an hour or two and in the end I turned it down. The thing is – I'm so dumb I'd surely be caught and I'm so chicken that long before they drove the wood splinters under my fingernails I would have ratted out all the neighbours too.

Incidentally, my wife thought – and still thinks – that we shoulda dun it.

Table Scrap: Still Neighbours Run Deep

There was a large market garden south of us on the four-lane to Winnipeg. In front of the house/yard complex was a sign, "Manitoba Potato Farmer of the Year." In fact there were three signs covering three consecutive years.

I mentioned this to Joe and Mary. They replied that he was a good farmer – after he got caught. (??)

For years the guy had made white lightning (vodka) from his potato crop. He had an insulated cube van to deliver his wholesale produce – corn, cabbage and such. The double-wall van body was insulated with moonshine.

On a trip to Kenora he stopped to gas up. He should have used self-serve. The pump attendant noticed a leaking spigot under the box, sniffed it and called the cops – busted!

Now he had to market his spuds legally. The folks around and about already knew he was a champion potato guy and now so did the government.

Table Scrap: Still Operators Run Deep, Run Silent and Run Big

I have a friend who told me this story years ago. He grew up in the Interlake country – Moonshine Central.

It was sort of a foreign owned operation, if you catch my drift. Local barns were used but others owned the stills. The farmer was well paid and when the inevitable bust came down he was supplied with top-of-the – line legal help. The first infraction was always a fine paid by the company. The operation moved to another barn.

I asked my friend if their barn was ever used. No, but his dad's ¾ ton GMC was part of the transportation fleet.

All "fleet" farmers parked their trucks outside with a full tank of gas and keys in the ignition. Once or twice a year, in the morning, the farmer would find his truck parked in its usual spot with two or three hundred miles on the clock. It would be low on gas but on the seat was an envelope with a wad of cash.

That operation ran as smooth as "smoke."

192

Joe and the Massey 26

Joe and Mary were entrepreneurship poster people, mixing under-the-table income with legitimate ventures.

Massey-Harris post-WWII had developed a self-propelled combine. I don't think they invented the concept but they down-sized it, making it applicable for smaller acreages. Most farms in Southern Ontario and parts of the West were still 80 to 320 acre parcels.

I think the first model was the 24, followed by 25. In 1950 they came out with the 26, slightly larger and a more efficient grain saver.

Joe was a forward thinker. His neighbours were still harvesting the old way – stook the grain sheaves and wait for the roving threshing machine, which might not arrive until after the first snowfall. Then the sheaves would have to be stacked to wait for spring. Fall income would be missed and rodents would extract their share over the winter. So Joe decided to go into custom harvesting. He needed two pieces of equipment. He already had a Cockshutt 20 tractor purchased in '49 with smoke cash. He also had the cash for an eight-foot pull-type swather ($600,) but he needed a bank loan for the combine. He went to the Massey dealer in Beausejour. There were two combines on the lot – a 25 and a 26. The 25 was priced at $2200 and the 26 at $2500. Joe wanted and needed the 26.

Now he went to the bank. The manager, as bank managers do, thought Joe was taking on more than he could handle. He would grudgingly back him on the 25 model, but Joe knew the 25 would not be adequate – back to the dealership. The Massey dealer knew Joe and unlike the city-slicker banker, he knew Joe was right. He told Joe to take the 26 home and pay for it when the crops came off.

That fall Joe ran that machine 24/7 for two months. With the last bushel of wheat in the last customer's bin he went to Beausejour and gave the dealer $2500 cash – the guy would not accept any interest.

Sidebar: Joe and Mary were the most remarkable couple I have ever known. To look t their farm you wouldn't know they owned any machinery. When machinery was bought a shed was built to house the purchase – one properly-sized building per unit. Joe told me the combine had only spent one night outside, when a gully-washer had caught him at sundown.

In 1987 the 26 still sat in its shed, unused now for years. Joe said maybe he should sell it, but who would want an old combine. I told a friend, retired on an acreage 30 miles north and he was interested. They put a new battery in the 26, fired 'er up and drove it home, where it took off 30 acres of barley/oats – original tires and original belts and not a whimper. Joe got $3000 for that 37-year-old combine.

Sidebar: These folks didn't care much for banks, but it was necessary to keep a bit on account for some expenses. One day when the 5-Star had been tipped more than usual he told us it was a 50-50 deal – a hundred grand in the bank and another hundred grand buried in a five gallon cream can! As the saying goes where there's fire there's smoke.

Sidebar: This I know is true because I read it in a newspaper in 1968.
Massey Harris sold a lot of tractors in England – they were popular over there. Canadian farmers realized that due to the mysterious ways of international trade they could buy

Masseys in England much cheaper. They got together as a group, and saved $2,000 per unit with transportation back to Canada factored in. Weird, huh?

I need a job so I try my luck at Manitoba Rolling Mills in Selkirk. They want a grade 12 certificate to pile steel and the high school in Trenton, Ontario where I had done my final year had burned down taking all records with it. (Not my fault, honest.) The mill believed me not – I guess they figured it was the equivalent of "The dog ate my homework."

So I spend the last few months on pogey, studying and writing my G.E.D. exam. (High School Equivalency.) I get good marks, too, but when I go back to the rolling mills they still won't hire me.

I tell my job counsellor "They think I'm too old."

"They can't ask that question," he says, "It's not on the application."

I point out that I last attended school in 1959. If they can't add 25 years to that and come up with 45 they would be too dumb to work for anyway.

Beggars can't be choosers and I'll try anything at least once, so I go to work at an auto body shop.

It's a small operation. Dad is mostly out of the shop now and the son is hands-on manager/top gun paint whiz. An older guy is our main source of "How do we do this?" and an 18-year-old is apprenticing. There is room for four units in the shop including the one in the paint booth. I've always been a car guy and despite the dust and paint fumes I enjoy it.

One day I help the older tech put a new rear quarter panel on a 1966 Mercury Parklane 2-door hardtop. Quarter panel my butt – the only larger piece of tin I've ever seen is on a barn roof.

We have a rusty old Econoline van sitting beside the shop. This is a practice project. From time-to-time we might have a couple of hours between jobs and the kid and I bring the van in to do some patching and filling. I think it is a weirdish sort of deal – it will eventually prove to be WEIRD for sure.

I am told the van owner is in no rush. He occasionally prepays six or eight hundred dollars and never complains if repairs go slowly. It is a mystery which super-sleuth will solve six months later.

Caution – this is <u>not</u> a feel-good story, but it is absolutely true. I will also be cautious. I don't want to be on this guy's top-ten hit list, if you catch my drift. I'll call him Q – you will soon see why.

One day a taxi pulls in, a guy steps out and has a conflab with the boss and his son. Money changes hands and the taxi leaves. The boss comes in with a thick wad of paper money - $800 in ones, threes and fives! We will continue to patch the van.

I ask the boss if he knows who the guy is and where does he live? He only knows the last name, Q – and a shiver runs up and down my knobby spine.

Two years ago a girl's body had been found in a ditch north of Winnipeg. Q was arrested, and appeared in court. He had no recollection of events, and in fact did not understand the charges or the court system. The judge remanded him and ordered a psychiatric evaluation.

194

He completely flummoxed the shrinks and was transported to the psych hospital in Selkirk to be held in secure (???) custody pending re-evaluation.

Every six months he reappeared in court, bamboozled the dummies and went back to secure (???) custody. Could this be the van guy?

I tell my suspicions to the boss and his son. They are skeptical – they have never heard of the Q back story. I am not completely convinced myself – surely it could not be the same guy!

A couple of weeks later a spiffy black-on-black-on-black Buick Grand National Turbo pulls up to the shop. The car's owner is proud of his baby and someone had put a parking lot dent in his door. While we do the minor repair I find out he is a guard at the psych joint. I tell him about Q and he is immediately alert.

He says security at the joint is lax. They are underfunded, undertrained and understaffed. Most patients are on mood-altering meds, easily hidden and easily marketed. He says visitors are popular and walk-abouts with residents are allowed. Meds change hands for peanuts – thus the wad of small bills.

Less than two weeks later a small item appears in the paper. Q is formally charged with the crime and subsequently convicted. The court had finally taken off their dunce caps.

I don't know what happened to the Econoline van, but I know my boss was never asked to return the $800.

A couple of years later I part ways with the body shop. The problem is that a body man has to be good and fast. I am neither with a capital "N." We part on good terms – I gave it my best shot.

Maybe I'll hit the bush again. Jim is back on his feet now, and he is as happy as a pig in poo. An old friend has bought up the lion's share of the stock (not hard at a penny per) and has re-installed Jim as prez. The stock has been newly underwritten, the treasury is full, and with a new board of directors the future looks rosy.

I visit Jim at his Winnipeg office and he is beaming. He tells me he can walk down the sidewalk again. I am a bit perplexed - what does he mean?

It seems that for the last couple of years Jim has been using a back lane to walk from his apartment to his office. He owes money at the bank, he has been unable to maintain regular payments, the bank is between his office and home, and the manager's office faces the street. Now, for the first time in years, Jim can walk past the bank and wave at the manager. Sometimes it's the little things that make life worth living.

So I help Jim out on occasion. One weekend he needs a hand staking a block in the Bird River area northeast of Lac du Bonnet. It is a rush job, so Jim charters a Jet Ranger and picks me up in my front yard. Sure impresses the hell out of my neighbours!

This particular staking job is one of several Jim has been doing lately in consultation with a rich widow. She loves dabbling in mining claims, and she is always the one who decides where to stake. Her system is very unconventional – yet effective. Get this – she lays a map out on the kitchen table, takes out her dear-departed's gold watch and chain, and dangles it over the map, running it back and forth, as if she is flying an airborne survey. If the watch starts to circle, she waits until it stops, sometimes moving it a bit before it settles down. That is the spot, and if the ground is open Jim stakes it. Jim has watched the process and he says it is downright eerie. Is it successful? I'll never know. Jim is out of the picture now, and besides, the whole Lac du Bonnet-Bird River-Bissett area is a rock hunter's paradise. You can knock a chunk off virtually at random and find something of interest.

One day we get a phone call from my cousin Charlie Jones (on my mother's side.) He is in Winnipeg and can he come out for a visit? Of course he can.

Charlie lived in Yellowknife NWT. He had his hoisting ticket and owned two draglines. He contracted mine and government jobs, hoisting steel and driving piles – he had a reputation as a hoister bar none.

Last time I had seen Charlie was more than ten years before when he dropped in at Vermilion Bay on his way back to New Brunswick. As the unofficial historian of our maritime family division, he always had good stories to tell. I enjoyed Charlie's visits.

So Charlie pulls into our driveway, steps out of his rental vehicle, and his left arm is in a sling! Right off the bat I know there is another good story coming.

He tells us he was in Yellowknife, dragging, hoisting and minding his own business, when the Federal Department of Public Works contacted him. Would Charlie be able to go to Baker Lake to drive some piles? Very interesting.

Check your atlas – Baker Lake is at least 600 miles east of Yellowknife and two hundred mile inland from the northwest end of Hudson's Bay. Charlie knows he can't possibly move his own stuff, but the government guy says there is a pile driver on site. A good-sized wharf had been built at Baker in the early 50s and the dragline was still there – parked and unused for more than 30 years.

Now Charlie knows a thing or two about life expectancy of machinery in the arid arctic environment. Things don't rust up there but they may deteriorate, so along with his tools he takes some spare belts and hoses. He also knows that bird nests and rodent vandalism will more likely be the issues he has to deal with.

At the Baker Lake townsite he is given a government pickup and hired hand. They motor two miles to the lake where the old Bucyrus-Erie dragline/pile pounder sits near the dock/wharf, appearing to be in good shape and un-vandalized. The cables are not rusted and as it had been parked on a cement pad the tracks are high and dry.

Like all draglines of its era there is a steel cabin housing the motor. The door is tight and there is no evidence of a rodent invasion. Charlie says it was quite roomy-it turned out to be short on elbow room.

He checks the exhaust stack for bird's nests, then all fluids are drained. Filters are changed and fluids replaced. New batteries are installed and when Charlie hits the starter the old girl fires right up, soon settling into a smooth purr. Even an old hand like Charlie is impressed – he says that after all those years there was not one hiccup. But after all, if the machine had been new when it was hauled into Baker it had only done one job, driving piles for one wharf.

Again, like most of these units in the mid-twentieth century, the engine rad was in the wall of the enclosure, driven by an electric motor and vee belt. Charlie's main concern is the

196

flexible coupling in the driveshaft leading to the hoist drums. He watches it closely for a while. Things seem to be copasetic so he turns left and sticks his elbow into the spinning fan!

He doesn't lose his arm but his elbow is smashed and an artery cut. With his right hand holding his tattered jacket sleeve tightly on the bleeding artery, he makes it down the ladder to the ground where his helper, seeing a bone sticking out, goes goofy and starts running in circles! Charlie tells him to settle down – he's not the injured one.

The kid is sent to the nursing station alone. Charlie, always cool under pressure, figures the kid may crash and burn. Charley also figures he can walk two miles if he has to.

The kid makes it. People arrive, a tourniquet and pressure bandages are installed, and an air evac, already warmed up, flies Charlie to Churchill. The doc in Churchill takes one look at it, sends him on to Winnipeg, and that's why he got to visit us. What a story – but you ain't heard nothin' yet.

You may ask, "Why put new posts in at a perfectly good old dock?" Well – we'll have to ask the Windsors.

Princess Anne had toured Canada the year before, including a stop in Yellowknife. Baker Lake was not on her itinerary but Anne was told they would really like to see her. She wanted to see Baker also, so she promised to return next year, which is this year, which she did.

Also – on this trip she would have time to tour the lake itself, thus the need for fresh pilings.

Charlie lost 50% of his elbow movement but he was still able to pull levers until he sold out and retired. I never did find out if he got an "Atta Boy" from England. I'm not a Royalty fan (book royalties aside) but I'll cut them some slack. They probably were never told of Charlie's sacrifice.

(Interesting fact: Baker Lake is Canada's geographical centre.)

Table Scrap

Where the Berens River empties into Lake Winnipeg stands an island, and on the island stands a fish processing plant. It is a fair-sized operation and is self-sufficient, having a large diesel generator, a huge freezer room and an ice-making plant. In the summer months a small management group lives on site and labour is supplied by locals just a short boat ride away.

The plant is fired up every spring, right after break-up. Local fishermen bring in their daily catch to be processed, frozen, and periodically shipped by lake to Selkirk. The complex is shut down in the fall – winter fish are easily transported to the main plant down south.

This enterprise and at least one other was once owned by an Icelandic family in Gimli. The Manitoba Government essentially nationalized the businesses, buying out all the Lake Winnipeg plants. As per usual they put their own staff in charge, needing no input from those who knew the ropes.

They fire up in the spring and although the big diesel runs well at idle it will not handle a load – no acceleration. Experts are brought in, expertly turned out in shiny new white hard hats and crisply laundered white smocks – and they don't have a clue. Other smocks arrive and they are also befuddled.

Walter C. (aka Bambi) has maintained these plants for years. They reluctantly call him in, greasy overalls included. Bambi looks the situation over for five whole minutes and then, as he had done every spring, he climbs up onto the roof and removes the bird nests from the twin exhaust stacks.

The following summer Jim has a job for me near Bissett. I throw my axe, pup tent, and Coleman stove into my old pickup and head 160 miles northeast, stopping in Lac du Bonnet to pick up a rented magnetometer. I pull into Bissett and grab a hamburger at the old San Antonio Hotel. One of the brothers still owns the place and he remembers me! I'm rather pleased; 25 years have passed since I last stepped through his door, and from the prosperous look of the place he must have had a customer or two. The man has a fantastic memory, I guess.

The town hasn't changed much. The hotel has added a 4-unit motel across the street, and just up the hill is a small grocery store with an attached restaurant.

After lunch I head on to check out the job. The 40-mile grid straddles the road about 15 miles east of Bissett on the way to Green Acres (Long Lake - Book One.)

This is cabin country now, and the old turkey track of 1963 is now a well-maintained gravel road. I find the line cutters' camp and check a couple of lines, and they are neatly cut and chained. Looks ok - the cutters must know their business.

I find a little park not far away – not much more than a couple of picnic tables, but it's neat and clean. I pitch my tent and establish a base station. Tomorrow I'll start reading the baseline.

(Before I left for Bissett Jim had told me that the line cutters had been on the job for more than three weeks and should be almost done. By the time I catch up to them the 40-mile grid will be completed.)

The next day I run into a line cutter half way down the baseline, stop for a chat and the news is not good. He is the only guy working. The rest left for Winnipeg 10 days ago and he hasn't seen them since. They are on a toot.

I am not pleased to say the least. Dammit, it's déjà vu all over again! The grid is half done and I have been hung out to dry. Well – never say whoa, etc. etc. Maybe the boys will sober up and return. Maybe the lone cutter can finish the grid on his own, so I carry on. As I travel back and forth setting up my base stations I pass the cutter occasionally and it doesn't take me long to figure out that if I wait for him to finish the job I'll be here until Christmas.

Let me tell you about Moose. He is a nice old guy from Bissett and is at least 60, and probably older. When the line cutting crew blew into town, Moose looked up the boss and offered his services. He likes the bush.

Moose is a one-man act. He works alone and cuts and chains at the same time. I've never seen anyone use his system, but it seems to work for Moose. He has 100' of light rope and when he starts his line he ties one end of the rope to a 3-pound rock which he plants on the baseline. He cuts until the rope plays out the full 100 feet, plants the station picket, pulls the rock to the picket and starts cutting again. It's definitely unconventional and Moose is pretty slow. In fact, he is very slow, and it's not entirely due to his system – Moose is an herbalist.

One rainy morning I go into town for breakfast and I ask the waitress if she knows George Williams (not his real name) and I draw a blank. Then I ask if she knows Moose, and of course she does, everybody knows Moose, Moose has lived here for years and years. Moose is locally famous, and during the conversation I find out that he actually is famous in the health food/holistic medicinal world. Moose collects plants, berries and roots; grinds, mixes and bottles them, and markets his concoctions world-wide on the internet. This nice old affable gent is an international entrepreneur. Still waters can sure run deep!

That afternoon I catch sight of Moose from a fair distance and watch him for a while. As he swings his axe he is scanning the underbrush, and from time to time he takes off his ever-present packsack, picks a few leaves and berries or digs up a couple of roots. Moose isn't here just to cut lines – the grid is his garden patch. I no longer worry about Moose and his lack of production. I think he is one cool gent.

A couple of days later I take a break. I'm almost finished the cut portion of the grid and I figure if I go home for a day or two the line cutters may show up.

My wife has a few days off so she returns to Bissett with me. She's never been involved with the bush other than berry picking and I figure its time she found out where the bear does his business. We get a room at the motel and the next day we head for the grid. She will carry the notebook and record my mag readings.

No line cutters have appeared yet and we have only 2 days' work ahead. My wife is a dandy notebook carrier and on the second day we come across a little creek in an open dry swamp. This had been a large beaver dam at one time, but the beavers had stripped the poplar edibles and had moved on years before. Only the creek remains and we hit it at a spot where it is 8 to 10 feet wide. It is quite deep here but I have crossed it before at narrow spots and I know the bottom is solid with no loonshit to worry about.

I have a station to read at the edge of the creek and I'm pretending to have some trouble taking the reading. Will this girl never learn? Once again she hits the bait.

She wants to know how we will cross, and I tell her it's easy – we will just wade. I'm still bogusly struggling to get my reading so I tell her to head over before me; but she'd better not get that notebook wet!

It's a hot day and she's half way across, up to her boobs in warm water and happily waving the notebook over her head. She is singing her own praises and bragging on how successful she is and climbs out the other side, soaking wet but proud, and now she wants to know how I'm going to cross without getting the mag wet. I say, "No problem," and I go up stream 40 feet to the old beaver dam hidden behind a line of willows, walk over and come back down the other side dry as a bone.

Well holy cow! The air turns blue! Every whiskey jack, raven and blue jay within half a mile takes flight and it's a good thing I can't hear what she thinks of me. I'm laughing so hard I have to sit down and it's a struggle to complete the line. It's tough to hold the mag still when you're laughing like crazy

(This was her final lesson. She has graduated on the honour roll and will never trust me again. The poor girl has been there and done that.)

Back at the baseline we run into Moose and he wonders if we had noticed the black bear who has been following us all morning. Well, that tears it. I can't chase that lady back into the bush with a stick, and I'm not too brave either. There is only half a loop of cut line left to survey anyway, so we check out of Bissett and head home.

The next day I finish plotting and take the maps in to Jim in Winnipeg. I tell Jim I can't do this anymore. Once again I have turned in an unsatisfactory, uncompleted survey and once again I have failed to turn a profit. I say goodbye to Jim – and I'll never again set foot in the bush. It's been a long slow process but I am finally weaned.

Chapter XII 1988 And Onward

Like so many folks do in tough times, we buy ourselves a job. A "For Sale" sign goes up at the hobby farm and Mr. Ryder rental truck takes us back to Ontario. We've leased a restaurant at Mine Centre.

The restaurant has been closed for a couple of years and we have a lot of work to do before opening. One day I mention to Joe, who owns the building, that I'm eager to see Harry Bell, who I am sure will be pleased as punch to hear the news. He tells me that Harry died last year. He had been out paddling on his beloved Bad Vermillion Lake, returned home, lay down for a nap and never woke up. Dammit! Another good man and good friend gone. I have to find a quiet spot to sit and ponder on the fates of life.

The restaurant, to everyone's surprise but ours, is successful. Mine Centre is pretty small, but the logging industry is booming. Also, some mining people are always kicking around, and a nearby First Nations community is very supportive. Added to the mix are long distance truckers. The Trans Canada is undergoing a lot of construction and delays are long and frequent. All truckers keep their ears on, and the word is out. They can leave the Trans Canada at Shabaqua Corners east of Atikokan, run Hwy 11 through Fort Frances, pick up Hwy 71, and rejoin the Trans Canada near Kenora. The route is only 30 miles longer, and besides, there is a good spot to eat at Mine Centre.

One day a trucker walks in looks around and says, "So this is the place." I ask him what he means, and he tells me that he was passing Barrie when he heard two guys bragging on us, so he just had to check us out. Barrie is only a few miles north of Toronto and more than 1000 miles away. Wow!

Our insurance agent stops in for coffee. Although he lives and works only 40 miles west of us, we don't see him often, and he has never had a meal here. It's just past lunchtime and I'm having a cup at the counter with him while a group of our First Nations customers are paying for their lunch and leaving. Our agent turns to me and, all smugly self-satisfied, he says, "Their money is as good as mine, eh?" Whoa! My feathers are ruffled.

"Not really," I retort, "Your money jingles, their money is paper with big numbers on the corners."

Jeez! People like this I can do without!

Cackleberries and Stovepiped Spuds

We are learning a lot about supplying a semi-isolated restaurant, and even more about human nature.

We shop wholesale at a national chain outlet 40 miles west in Fort Frances, and also from a Fort Frances family-owned food services distributor. Ernie and Lois Thompson (the family folks) have been in business since I was a kid. They are very much hands-on service oriented, and have an ironclad reputation for honesty. The two outfits are our main suppliers – our criteria are quality and price.

Our two main problems are potatoes and eggs. Let's deal with the egg problem first.

We do a $2.99 breakfast – two eggs, bacon, pan fries and toasted homemade bread. Pan fries are no problem – any potato pre-boiled and diced makes decent pan fries. Hotel-sliced bacon is cheaper at the wholesale: eggs are cheaper from Ernie and Lois, with one little drawback. Their eggs are locally sourced from a farm in the Rainy River Valley, and the quality of their product tends to be uneven. My wife grew up next door to a big egg producer, and she knows a good egg when she sees it.

We close the doors only twice a year – one day off at Christmas and a four-day shutdown on Easter weekend. On Good Friday we bring in a couple of local recruits to help scrub, sweep, dust and sterilize, and by 4pm the place is squeaky clean. We lock the doors, jump into our van, and head 300 miles west to spend Easter with my wife's parents in Manitoba. On Easter Monday we head back after stopping next door at a Hutterite colony. We pile 12 cases of top-quality grade A eggs in the back – no egg problems for two weeks.

Shortly thereafter we are visited by another valley resident who is now running once a week to Atikokan delivering eggs to people such as ourselves. Do we want fresh eggs? We check out the product. The cardboard cases containing 15 dozen eggs are sealed with brown tape and stamped on the tape is "Ridgeland Colony," our supplier of choice right next to Mom and Dad in Manitoba – we jump at the chance.

For the next ten months we have great eggs dropped once a week at our door. But bango – the quality disappears! Now some eggs are cracked and others are double-yolked. The doubles are a bonus for our hard-working loggers, but is a definite no-no in the restaurant trade. If he gets a double-yolker, why can't I?

Sometimes, when a normal-looking egg is put on the grill it spreads out as big as a pancake, others sit up like they are soft boiled. Our eggs are still arriving in Ridgeland Colony boxes, but the seal is broken. Easter is just around the corner. Super Sleuth takes his cape out of storage.

We tell my wife's father about our eggs, and he can't believe that the colony is at fault. Dad has dogs and cats, and he knows that an occasional egg is a treat for them, and also has health benefits. He buys rejects from the colony at two dollars a case – hmmm. On Saturday we pay a visit to the egg man next door.

I find him in the grading station attached to the laying-hen barn. Here the eggs are washed, graded, and packed into cases to be shipped to the Egg Marketing Board in Winnipeg.

The egg man is an older, pleasant, Santa Claus kind of gent. I feel a little hesitant about accusing him of skulduggery, but I forge ahead. He is shocked, insulted, and is not shy at all about defending his operation. I am going to get an egg-grading tutorial.

He points to a wall where full cases of eggs await shipment. "Young fellow," he says, (I am 50) "There are 300 cases going out next Monday, and we ship three times a week. At any time, an inspector can come in and ask for three cases to be checked at random. Every egg in the case is looked at, and if there are three cracked eggs in any case, I have to destroy the rest of the three hundred! Why would I take that chance?"

Now I learn about the Haugh test. The inspector picks three eggs from the cases and cracks them open. The height of the egg is measured: Too high – stale egg: too low – stale egg: three Haugh failures in a case and the whole stack has to be destroyed.

The gent is cooling down now, so I tell what we are getting supplied with, and who our supplier is.

"Sonny," he says, "X has been buying a hundred cases a week from me for almost a year, but I haven't seen him for two months. I think you can figure this out for yourself."

Well, I sure can. I sincerely apologize, buy my twelve cases, and go back to the father-in-law's with the story. I will also have a heart-to-heart with X next week.

You've also figured it out now, haven't you? X has found someone to buy rejects from, and is selling us cases of two-dollar eggs for $21. He has saved up Ridgeland boxes to repack them. What a crook!

The next week X comes in with a shipment of crap and I meet him at the door. He turns bright red and beetles back to his cube van.

The following week we are delivered eggs in sealed Ridgeland Colony cases, but X stays outside. One of his daughters brings the eggs in.

Egg-gate is resolved, but we have similar issues with our potatoes. We cut our own fries, and serve them skins on – fully dressed, you might say. So we need washed, uniform-sized, unblemished potatoes. They can also be baked, or easily peeled to be boiled and mashed. We don't want to mess around with different grades – one size fits all.

We like to buy local, and we get our potatoes from Ernie and Lois, who get them from a valley farmer. After delivery we open bags to check for quality. The spuds look good, and as potato bags have sort of an open weave, we can even check those farther down the bag, but as we use the potatoes, we start to find duds. Some have been sliced by the digger, and others have spots of rot, are not well washed, or are pretty small.

The next time Ernie comes I tell him his potatoes have been stovepiped. Ernie gives me a quizzical look – it's my turn to give a tutorial.

I was born in New Brunswick in a small town in the Saint John River Valley, not far from Florenceville, where the McCain frozen food empire has their head office. This was and is potato country, and although I was hauled west after WWII, my mom's family, all United Empire Loyalists, still have a presence in the area. Therefore – I know how to stovepipe potatoes.

It's very easy to do. You have a potato sack and a 3-foot length of 10" stovepipe. You centre the pipe in the bag, fill the bag around the outside of the pipe with good potatoes, fill the pipe with cull spuds, pull the stovepipe, and complete the fraud with a layer of good potatoes. Now you tie the bag, and voila – stovepiped potatoes.

Ernie is sceptical, but interested. On his next trip by our place, he stops in to tell us he had dumped three bags of Mr. Y's potatoes on the storeroom floor. Just as I said – good potatoes on the outside, trash on the inside.

Almost immediately our potato quality zooms upward.

Table Scrap: It's What You Know <u>and</u> Who You Know.
Back in 1960 when I was on Setting Lake, I sent my mom $300 to help her out for a trip back to New Brunswick to visit family – her first visit since 1946. Before she left, Dad told her there would be two hundred dollars in the bank at all times, should she need extra money. Mom had a sister in Florenceville, and before the return trip, she realized she would have to tap the $200. The bank in Florenceville would not cash her check without a long distance phone call, and if you knew my mom, you would know that she would never part with five dollars for a phone call, if she could help it. She walked out of the bank, and right across the street was the head office of McCain Foods. My mom had gone to school with Gilbert McCain, and after graduating as a registered nurse, one of her first jobs was as a personal

nurse to Mrs. McCain, Gilbert's mom. She walked across to the office, asked to see Gilbert McCain, wondering if he would remember her. It had been forty years, but Gilbert sure did remember! After chatting for a while, mom explained the check cashing problem, and wondered if he could help her out.

"Come with me, young lady," and back to the bank they went. The bank manager was only too pleased to cash a personal check for Mrs. Durnin.

Never Hire a Direction-challenged Prospector as a Tour Guide.

There are two main forestry access roads near Mine Centre. Both run north off Hwy 11 Five miles west of us is the Turtle River road, and five miles east of town is the Manion Lake road. For more than a year we have listened to the wood haulers discuss these roads – it's time to check them out.

Let me digress a bit: In a small town it's tough to find help. Those who are retired don't need a job, and everyone else is working; the employment rate in Mine Centre is 100%. We have one full-time waitress and she's a dandy, but Debbie has her own family to care for, and she works a 40-hour week on day shift. We've hired two students who work evenings and weekends; Anna is sharp as a tack and Danny is quite a kid. He starts out washing dishes but he's an ambitious rascal. He keeps an eye on my wife, always curious, and only too willing to grab a spatula. He has his own progression chart, and one day I go into the kitchen and my wife is up to her elbows in soapy water and Danny is flipping burgers, whistling a happy tune, and barking orders at his dishwasher/assistant. My wife is pretty sure there is something wrong with this picture, but she's smiling.

(Danny went on to school back east, paying his own way using his culinary skills. I think he may be a lawyer now. I hope he's an honest one.)

So one day in June, Danny gets off the school bus and finds himself in charge. The wife and I are going to check out the wood haul.

I've been told that we can do a loop, so we head up the Manion Lake road; the plan being to drive north 40 miles or so, cross over and pick up the north end of the Turtle River road. We chug along enjoying the scenery and waving to our log-haul friends and customers as we meet them along the way. We pass the airport and I'm looking for the cut off at the north end of the loop. If we continue on Manion we will intersect the Bending Lake road which runs from Raleigh Lake (west of Ignace) to Atikokan, and we don't want to go that far today.

Sidebar: They call it the airport: A mile-long straight stretch of the Manion Lake road has been widened and improved to make a decent, flat runway. Once a year the gates are closed at each end and a crop duster comes in to spray Agent Orange, reloading and refuelling at the strip. The idea is to defoliate the broadleaf saplings in the replanted areas, giving the young conifers access to sunlight. It's supposed to be harmless to the wildlife, but the absence of birds following the Agent Orange application belies that theory. Dead sparrows tell no tales.

I make one false start on a tote road, stop a wood truck and he tells me our road is two miles further up. He tells me that after 15 miles west on the crossroad I will pick up the Eltrut Lake road, which will lead us to the Turtle. Well, all my life I've had a bad case of A.D.D. and the Eltrut info blows over the top of my head unheard.

We pick up the crossroad and motor west. We are in no hurry. In June the evenings are long and the restaurant closes at 10 p.m. Danny is supposed to go home at 9:30 – no problem.

We cross a little bridge and wave to a group of blueberry pickers camped on the riverbank. (There is a good dollar to be made from wild blueberries, and many families pick in late June and July). A couple of miles further along we reach a fork in the road, and I stop to study on the situation. The road to the left seems to go in the proper direction, but it's not too promising. There is a grass strip down the centre and it hasn't seen much recent traffic. The road to the right is much better looking – it's nice and wide and well gravelled. The left road has a sign – "Eltrut Lake Road," it says, but that means nothing to me. My brain cramps and I turn right.

It's a nice warm day but completely overcast. The sun is up there somewhere, but we can't see it. We mosey along, pick a few berries and toss a line into roadside lakes snagging the odd tree branch. My wife is getting edgy, and has the temerity to mention the L-word (lost), "Ha ha." I scoff, "You can't lose the old prospector."

The gall of this woman! I don't mention the hip wader incident at Black Beaver in 1969. That's not something I brag about, but I'm getting a little worried myself. I've seen the south end of the Turtle River road and it's pretty wide, and this sucker is deteriorating as we drive onward. I'm toying with the idea of turning back, when we round a bend and a half a mile ahead transports are flashing by on a major highway. My relief is short lived. We should be crossing a railway, and no such thing exists. Holy cow! We've come out at Borup's Corners, no more than 50 miles east of Dryden, on the Trans-Canada highway! We have run the Snake River Road, which hasn't seen a stick of wood for 5 years!

Heigh-di-ho, which way do we go? We could backtrack, but it will be slow traveling, and it is almost 8 pm! We have a long way to go and a short time to get there. Plan 'B' it is!

So it is wide-open throttle eastward 20 miles to the Bending Lake road, and on 15 miles southeast to the top end of Manion. When we reach the crossroad we pause for a minute. We could simply continue down Manion and return with our tail between our legs, but all the ponies in the Mine Centre corral are pretty swift. They'll know we got lost, and we don't want them to laugh at us. We've made pretty good time so far, so we decide to complete the loop as advertised.

Now it's westward again on the crossroad with our Safari van drifting on every corner. We bail past the berry pickers trailing a mile-long rooster tail of dust, and they all stand gaping (I guess they are still telling their grandkids about those crazy people who only travel in one direction.) We hit Eltrut and mow 10 miles of centre line grass, hoping there are no rocks lurking – the Safari lacks a stump-pan.

We hit the top end of Turtle, and this is more like it. The road soon broadens and we are making great time. Scenery be damned – it flashes by unnoticed. We catch some air at the railway crossing, turn left to Mine Centre and slide into our parking lot á la "Hollywood Hell Drivers," (You know the drill – hit the parking brake, kick her sideways, and coolly step out the open door as the Safari settles back on all fours.)

It's 9:30 on the nose and we've completed our planned loop and have flown under the radar – NOT!

Danny meets us at the door laughing like crazy. The darn log haulers have been tracking our progress all afternoon. Their radios have been hot, and from time to time they stop in to

205

update Danny on our progress. He might as well have been sticking pins in a war room wall map. Under the radar my butt!

It takes more than a few days for the giggling to subside, but it's good for business. Everyone around wants to stop in, some to have a look at these fools, and all with advice and opinions: I'm Marco Polo, I'm Wrong Way Corrigan, my wife should hide my keys, I should pack a lunch when I go to town for supplies, and on, and on.

Eventually some poor guy lays a load of wood on its side and I'm off the hook. It's his turn to pay the piper.

Attitude Adjustment Part Deux
We run the Mine Centre restaurant for two years. For the first year-and-a-half we are open 24 hours from 4 am Monday to 10 pm Sunday.

There is more than one reason for doing so. Within a month after opening we realize that the kitchen is just too small. Although the place seats only 32, it is often full at lunch and supper-time. Breakfast can be busy also, and throughout the day the joint is rarely empty. My wife bakes internationally-famous cream pies and homemade bread, and we do not use rolled turkey or pre-cooked beef – everything is as fresh as possible.

The second reason is that I am the world's worst waiter – I'm sure I wouldn't last half a shift at Fawlty Towers.

So I run the night shift from 10 pm to 6 am. Few customers come in, and I can pour coffee and put a piece of pie on a plate. When prodded I can even flip a burger or make a decent Denver (western omelette) if I can't talk the guy out of it.

At night I cook big stuff to take the pressure off the kitchen. One night each week there will be a 25lb inside-round roast of beef in the oven. The next night I run half the roast through the slicer and package individual portions for hot or cold beef sandwiches.

Another night I will boil a 35 lb turkey and strain the stock for my wife – she makes great homemade soup. Then the following night I will strip the turkey and package individual portions a la beef slices.

At slack times I do the books at my office cubicle in a corner booth. Twice a week I sleep two hours, arise at 10 am and drive to Fort Frances to the wholesalers, returning at 2 pm to crash for a few more hours.

It's a strange existence – I'm always tired and stressed, and when given a chance I look for sympathy.

One of my regular customers is Daryl Romyn, a down-homer Valley Boy. He runs a chip truck between the Sapawe sawmill (not far east of Atikokan) and the Fort Frances pulp mill. He prefers to run nights and leaves the day run to a hired driver. Daryl always stops for coffee and a chat, and I really like him.

But Daryl has been getting a little cool lately – what is wrong?

He stops in one slow night and I have a coffee with him. I have slipped into a pattern with Daryl and have been complaining far too much about my miserable life. He gets up and tosses a buck on the counter, saying, "I don't want to sit here and listen to you whine."

I am shocked, astounded and uncharacteristically up to speed instantly – of course, he is absolutely right. I adjust my attitude and I feel better.

Daryl stays away two nights and returns to find a more cheerful conversationalist.

Thank you Daryl, and other customers should thank you also.

Coffee Cups and Cleanliness

The main protagonist will be called "G." This is to avoid embarrassing him and his family. The thing is – although he is a great guy, he tends to be a bit messy.

He drove night shift for another chip hauler, and always stopped for a coffee. It was a pit stop only – G was a road warrior.

He had a travelling cup and would set it on the counter with a dollar beside it. He would then hit the washroom while I filled his cup with black coffee, and he would pick it up on his way out.

The cup was once white and orange with a coffee chain logo on either side. Now, it was mostly black and sticky – very sticky – I handled it with the tips of my fingers.

So it came to pass that one night I had been washing dishes when G came in. I had a sink full of hot, soapy water, so I decided to do him a favor, and I washed the cup – no easy task.

He was waiting impatiently when I came out of the kitchen and filled a nice, shiny cup. He did not thank me – in fact, he seemed sort of miffed.

The next night he came in cupless and took a large styrofoam. I asked him where his cup was.

"Bob, "He says, 'You just had to wash the dam thing! When I got out of the truck it slipped off the seat, and I ran over it!"

Maybe I should have replaced the cup. He bought another and carefully cured it, until it became as sticky as the original.

Side bar: Coffee cups should be cured, like a good pipe. It takes a few cups of black coffee to do so, but once burnt in, the coffee tastes better. Wash the outside only to clean up drool – never wash the inside.

I use a black cup – once cured, light that enters therein cannot escape. The cup swallows it like a black hole in space. I forbid my wife to wash my cup – she ignores me.

Economics and Big-versus-Little Government.

One summer I have a gent who comes in every morning. He is Danish and is an electrical engineer. The paper mill has two power dams on the Seine River and this summer they are tuning up the generators. He stops in early and if I am not busy, we chat – he is highly intelligent, very interested in all things local, and very interesting to talk (listen) to.

I'm usually circumspect about ringing stuff into the till when a stranger is about (shades of the country store in North Carolina – one can never tell when "revenoors" drop in.)

He has coffee only one day and when I go to ring it in he interrupts me. "Stop that!" he says. "If you pay tax it only encourages them to spend it. It is the duty of all good citizens to pay as little tax as possible." I get the point.

One morning the subject of governmental centralization/ decentralization comes up. The sub-topic is welfare.

Welfare is needed, he opined, but centralization leads to welfare abuse, as Denmark found when they formed one central welfare department.

He said local is better. Locals know if someone is double dipping, and when Denmark decentralized, welfare abuse virtually disappeared.

Another point well taken. I agree wholeheartedly with him as I slip his dollar into my jeans.

Attitude Adjustment Part Three

One of my most valued employees is Trapper, our Blue Merle Collie. He has been our protector for years and now adds parking lot attendant to his portfolio.

He does fine until the truckers find out about us. The chip haulers pull "B" trains and the loggers run five-axle trailers. Now Trapper has fourteen or sixteen tires to mark and his ink supply can't handle it.

It's kinda funny to watch him ponder on his problem. Finally, he decides to mark one tire on each unit, but we still have to keep his water bowl full.

Early one spring morning I hear Trapper growling so I walk outside to check. It's Sunday morning – no traffic yet – and just across the road two big timber wolves are pacing on the highway shoulder. They are a tough looking pair, each one twice as big as Trapper but our boy is never one to back down. What he doesn't know is that the rest of the pack is hiding in the bush. If the two lobos can entice him to follow, the rest will close in behind.

I call Trapper in and give him his reward – one of my world-famous Doggy Denvers.

Dine & Dash – Part I

It's March 15 and something is going on. Loads of wood are streaming by on the highway – many more than usual, and some of the trucks are not often seen in these parts – they are guys who haul from the west end of the district.

When I come on shift the boys are still at it. A returning empty stops for a bite and I ask "What's up?"

He tells me that the mill woodyard will close and lock the gates tomorrow morning at 8 am. The issue is that spring break-up half-load restrictions take effect at that time and the mill will not take any more wood until restrictions are lifted. The guy they are hauling for is a good joe and will often neglect his own haul to help out someone with a mechanical or financial problem. He has 2000 cords sitting at landings on the Manion Lake road. Every cord left in the bush can be hauled after break-up - if the road to the landing is passable. Also – undelivered wood that is carried over will cut into the summer contract. It must be hauled now!

The radios have been humming all day. Truckers from all over the district have come eastward instead of going home.

At 2 am the last loads come down Manion and everyone stops in for pie and coffee. The parking lot and the restaurant is full and Trapper is continually refuelling at his water bowl.

It is a happy party atmosphere and it is a good thing my wife had baked six pies. I am as busy as a one-armed waiter

A civilian car pulls in – young fellas I have never seen before. One comes in, sees that there is some pie left, takes three styrofoam cups of coffee and a "to-go" slice of pie. He says he has to go out to the car for more money.

"They're leaving!" someone shouts, and sure enough – taillights are exiting the lot.

I AM PISSED.

I run outside and grab a baseball-sized rock. I do my Sandy Koufax imitation, but the rock just bounces off the trunk of an old Monte Carlo as it heads west.

Back into the restaurant I go and everyone is laughing like crazy – especially Orval Woolsey. (I hope you're reading this, Orval, you jerk.) He is absolutely gasping and falls out of his booth.

I fail to see the humour in the situation. I phone the cops and Dave S. bless his heart, stops the guy just west of the Fort.

The guy says it was an innocent mistake. Dave radios dispatch, dispatch calls me, Dave collects $6.50, and drives 35 miles to bring me the cash – good cop.

Dine & Dash Part II – the Sequel

This time it's a quiet summer afternoon. Returning from my wholesale trip I meet a cyclist halfway to Mine Centre. He is obviously on a cross-Canada tour with stuff rolled up on the rear fender and bulging saddlebags. Traffic is light, but I slow down anyway, give him a friendly wave and get one in return.

At the restaurant I am greeted by Debbie and my wife. Did I meet a guy on a bike? Yes I did. Well, he did not pay for his dinner.

Crafty guy – he ordered a full pork chop dinner, not a cheap item in 1989. He heard Anna talking with my wife discussing her afternoon plans – Debbie was relieving Anna at 2 pm.

Debbie came in and the guy left, telling Debbie he had paid the other girl. They called Anna – no, he didn't pay her – what to do?

Phone the cops, of course – which I did.

A young constable stopped the biker at almost the same spot Dave S. had apprehended the Monte Carlo. The guy said he had paid. Dispatch called me – it was "He said, we said."

And they let the jerk go.

One only successful dine and dash in two years and he escaped on a bicycle on a semi-lonely highway in the Canadian Shield.

There are Good Customers and Real Good Customers

Jack Hedman drives out from Fort Frances every day to teach a course at the Learning Centre at the Seine River Reserve. On his return he always stops in for a 3 pm lunch – he loves my wife's homemade soup and he always sits at a table by a front window. It's a real nice day in September and in rural Canada September is fly time. They hunker down overnight and when the day warms up they like the sun. The restaurant door opens and flies fly in. The door closes, and flies go to the window, but can't get out.

The staff is always diligent and keeps them honest, but today it's Monday and I am staff – need I say more?

Jack crooks his finger at me – "Waiter, there's a fly in my soup!"

I hustle over discombobulated, and Jack is laughing like crazy. "Look at the size of that rascal!" he chortles, "He just about capsized the bowl when he hit the drink!"

I am mortified – I will get him a fresh bowl. Jack will have none of that. He fishes out a huge blue-bottle and finishes his soup. (Jack is a down-to-earth country boy.)

I offer him a free pass, but he refuses that also. "I've been waiting all my life for the opportunity to holler, "Waiter! Waiter! There's a fly in my soup!"

101 Ways to Lose your Wife - #57

I have a regular bi-weekly middle-of-the-night customer. He hauls bread from Thunder Bay to Ignace, Dryden, Kenora and Fort Frances, and completes the loop on our Hwy 11, stopping at Sapawe to pick up a back-haul load of lumber. His wife often runs with him and it is their habit to stop at our joint around 3 am for breakfast and/or coffee.

So one night I get a phone call at 2 am. It is the wife, and she asks me if X is there yet. I respond in the negative, and I am perplexed – is she keeping tabs on him?

When she explains the deal I can't help laughing and she is laughing also.

He drops his Fort Frances bread at a storage trailer behind a 24-hour service station, unloading the bread while she sleeps in the bunk. She wakes up to heed a nature call, goes into the service station and hears him pulling out. With no local cell phone service she cannot contact him – I am to send him back.

The guy pulls in, and I can not help myself – I have to play the straight man.

I stand by the truck door, coolly and ostentatiously enjoying a smoke, while he makes some notes in his log book. Then he turns, parts the bunk curtains, swings back with eyes now wide open, and rolls down the window.

"Oh, yeah," says I, flipping my butt onto the gravel, "She's still on the throne forty miles back."

Well, he is _not_ amused. He is ready to jam some gears, but I tell him to cool down. "Take my car," which he does.

He returns with his lady and _she_ comes in alone to thank me – he never stops again. Well too bad, lad, pee on you if you can't take a joke.

In 1990 with more than a little help from our friends we buy a little country convenience store/gas bar and once again we turn a sow's ear into silk purse. The place was pretty much written off locally as a dead horse but within less than a year we can't run people off with a baseball bat, (Really! My wife always keeps a Louisville Slugger handy and she's not afraid to use it.)

Table scrap

When we purchased the corner convenience store/gas bar we were new at the game, and we thought we might expand into frozen meats: the idea being to have a few steaks and such available for folks planning an impromptu barbeque or whatever. We don't know how to work this out, so we contact a local meat supplier for adviceThe head honcho comes around with a guy he introduces as his sales rep, and we sit at our kitchen table for a discussion. We want to learn the pros and cons of selling meat, and most importantly, can we expect a decent markup while holding the retail price at a reasonable level? Now I don't think this is rocket surgery, but you'd never know it by these two guys – they talk in circles.

Finally, I've had enough double talk and I lay it out for them. What will our input cost be? I can do the math myself – will there be room for profit? The answer is simple, but shocking. They will supply us with Grade D beef and we will market it as Grade A, thus making our profit. I can't believe I heard that, and I don't trust myself, nor can I think of an appropriate response. I walk away leaving my wife to usher them out the door. Sheesh! (Incidentally – those guys are no longer in business.)

In 1993 we convert the living quarters into a restaurant and build an addition to seat 24 customers. As an added attraction to what is already a mostly enjoyable existence is the fact that three old friends are in the area. Don, who once walked five miles to join me for lunch at Smooth Rock Falls in 1966, has been living in Fort Frances for years, raising a fine family and he is still in the bush. My old friend Sam is retired in Fort Frances winding up his diamond drilling business and N.B.U. Jack is living quite comfortably 20 miles north, enjoying his well-earned wealth, still prospecting and occasionally doing some contract staking.

From time-to-time Sam and Jack come to the store, we all pile into Sam's car, motor out to Don's project and check out the core shack and just generally snoop around. (You understand; this is quite a privilege; the courtesy is extended to us because of our connection to Sam. He has known the president of the junior company for years). We return to the restaurant, have a late lunch and tell each other lies. Sam and Jack then go back home and I go back to work.

My wife and I have one shift to cover and one shift only. Seven days a week we wake up at 5 am, open the store and restaurant at 6 am, close at 10 pm, cash out, and go home to bed. Life is pretty good, and with Sam, Jack and Don nearby I can still keep a vicarious connection to the bush life.

It's not all nose to the grindstone. My wife's son and daughter-in-law are involved in the restaurant end of the business and our daughter Candy is our mainstay in the store. Candy is Miss Personality Plus. People come in just to talk and joke with her, and of course they have to buy something. If you supply entertainment there is a cover charge, right? We wish we could clone that girl.

Anyway, every Friday evening from mid-May to early September we turn the restaurant over to Gary, the store to Candy or Donna, and at seven o'clock it's off to the races. We are big time fans, and Emo, six miles west, has a 3/8-mile dirt track which we consider to be one of the best in the area – and by that we mean for miles and miles, including northern Minnesota. With three classes (hobby, street and modified) the entertainment can't be beat. We sponsor more than one car – there's not much money to be made in local dirt track racing and these boys are mostly working stiffs just like us. Our sponsorship is no big deal really, the odd 100-dollar cheque, and to help out a few guys we throw in 20 bucks worth of tow gas every race night. The problem is, that despite our objections the buggers keep painting our name on the side of their cars, so there are six or seven stocks tooling around the track every night with June's Place oh-so-visible. We owe a ton of money to the bank, and the manager, who is also a stock car fan, sits four rows below us in the grandstand. Every Friday night I watch his ears turn red as the announcer brags on us. I try to make myself as small as possible and I sure wish the announcer would clam up a bit. The manager never does say anything, though. He's a pretty good guy.

We have concocted a plan regarding our Friday night return from the races. When the last chequered flag has dropped everyone leaves at the same time (naturally,) and a steady stream of traffic heads east to Fort Frances. Our store is on the north side of the road and on a curve, and we always stop in to check on security. On most nights if the races have not had too many restarts, we can get back before closing time. The trouble is, we turn left into our place, and if we have to stop for oncoming vehicles, the line behind us will accordion very quickly, creating a very dangerous situation.

Therefore, our plan is thus: We park the van for a quick getaway, bolt from the grandstand and boot it down the road, and when I say "boot it" I'm not fooling. After three hours of watching dirt track racing everyone is charged up and folks behind us don't dawdle.

I round a curve at 75 mph keeping the nearest headlights a mile behind, and holy cow! We meet a black and white! I lift the throttle but of course it's too late now. I can't believe it. That sucker does a highway turnaround and in less than five seconds he's on my back bumper and the lights are flashing. I pull over, and when I see him in my mirror walking up to the van my heart drops. It's Tom. I know him a bit, both personally and by word of mouth (we know all the cops.) He is known as a fair man but he's sort of a take-no-prisoners type of guy, if you get my drift.

I roll down my window and when Tom sees me he jumps back a little. "Bob!" he says, "What in the hell are you doing?"

I give him a quick précis of my reason for speeding, he looks back at the oncoming line of headlights, slaps my door and says "Get cracking!" I peel off into our parking lot and an uninterrupted convoy of vehicles streams past. Don't ever tell me there is no such thing as a decent cop. I love that man.

Donna is another valued employee. She is a single mother with three boys, and although she can sit at home and receive mother's allowance she would rather work – even at minimum wage. The irony is, mother's allowance pays X, we pay Y. The welfare dept. subtracts Y from X = Z. There is no financial gain for Donna. She works for personal pride.

A suit from Dryden drops in. He represents a government program that encourages small business to upgrade the skills of entry-level wage earners. If we give Donna a raise, some training and more responsibility, they will pay half her salary for a year. I call Donna in and the three of us go over the program's high points. She jumps at the chance. It's still no financial advantage to her, X-Y still = Z, but her job will be more interesting.

Donna starts to handle stock control, learns to cash out, gets to sweep the floor twice a day (instead of once) and we can leave her in charge in our absence, knowing she will do a good, honest job. It's a win-win deal.

The suit comes back every three months, picks up a copy of Donna's payroll record and my detailed report on her progress, gets Donna's initials on both, writes us a cheque and leaves. At the end of the fourth visit he gives Donna a bogus looking diploma, shakes hands all around and leaves, forgetting my report on the desk. I recall my EBAU experience in 1972 and wonder if this guy gives a crap, but what the hey, I've gained $5 per hour x 2000 hours, Donna has blossomed into a well-rounded employee, and everything is cool.

Not long afterwards my wife hires a lady from Fort Frances, and she has a slightly different story for us.

Her previous employer had put her on the program also, but his method was a little more creative than ours. He doesn't give her a raise, but he keeps a second payroll record showing an increase, and his progress report is pure fiction. When the suit visits she dutifully does the initialling, nods her head if suit asks a question (happens seldom,) and has to do so or she will be canned. When the program ends her boss has pocketed 10 grand and she gets canned anyway. Did the suit care? I think not. He was just covering his own butt a la EBAU.

Same song, different verse: A customer I have known since high school tells me this story.

He stops often at a certain establishment and on slow days the owner joins him for a bit of BS. One of the waitresses seems really nice, always friendly, efficient and hard-working. For some reason the owner takes a dislike to her and begins criticizing her and chewing her out in front of customers. She can do nothing right! My friend can't understand how she suddenly became such a poor worker, and he remarks on this to the owner. The guy's answer? The girl has completed the suit's program. If the owner can bully her into quitting, he can hire a new victim and go back on the taxpayer's tit. What a jerk!

But there's one consolation. My buddy never sets foot in that place again.

We take a rare afternoon off and decide to tour some new country. I'm sure you know the drill by now – leave the pavement behind. We make the usual false starts, hitting dead ends and washouts, but that's the whole point – how far can you get into the bush, and still get out? And besides – we are looking for the Northwest Passage.

Some time ago a road had been built in an arc between the Splitrock area and Highway 71. It had seen the log haul for a few years, and is now in a temporary hiatus – no longer maintained – just waiting for some new timber limit activity. I am told hunters use it every fall – this should be fun!

About five miles in on a slowly deteriorating track, we cross a large culvert over a narrow spot in a huge beaver pond, park the van, and walk back to enjoy nature. Some beavers on the day shift (probably salaried personnel) are minding their own dam business. Some ducks are feeding, and a blue heron is shopping for dinner. It is so quiet and peaceful with the nearest house 12 miles behind us.

As we stand enjoying nature, I notice the tire tracks on the loose sand/gravel over the culvert. I chuckle and poke my wife in the ribs. "Look at that, honey. Some dumb jerk has been driving on a flat tire!"

No sooner are the words spoken, than I realize we are the first to use this road since last fall. Up the hill our van sits, a little lopsided. So DJ changes the tire.

It's our first flat on our Safari, so, of course, the spare is one of those little 2-ply pretend tires. We drive 50 miles home, very carefully, stopping often to let that little so-called "tire" cool off.

The next day a proper spare replaces the donut.

Tables Scrap

In 1984 an old Manitoba farmer picked up a new full-sized Pontiac from a long-standing family-owned dealership. He drove home 50 miles, proudly showed the car to friends and family, and when they lifted the trunk lid, there lay a make-believe spare. Straight to the phone – the farmer wants to speak to the owner – right now!

"Sonny," he says, "I bought my first Pontiac from your granddaddy in 1931, and I've bought them from your dad, and I've bought them from you, and if there's not a real tire in that trunk tomorrow, I'll never deal with you again!" (This was an 88-year-old guy, How many more Pontiacs would be in his future?)

The next morning there was a new wheel and full-sized tire in the trunk – overnight express!

So, in 1993 we are employing ten people and grossing 1.7 million. One night in March of 1994 the damned place catches fire! The culprits are a tired freezer compressor and a staple through an electrical wire.

Had it burned to the ground things may have turned out differently, but although the fire destroys the centre half of the structure from the floor through the roof, the walls are still standing. Every cooler, appliance, and piece of furniture is damaged by heat and smoke. The stock is worthless and the office is a total wipeout. We consider it a 100% loss. The insurance company disagrees, they call it a 50% fire. The battle is on, but they have the big guns. We have a pointy stick.

We are behind from day one. First of all, we have never had a fire before, but they deal with yahoos like us every day. Secondly, we fail to adjust our way of handling people. For years we have operated on one basic principle: the customer pays for a service, and we do our best to satisfy said customer. Now, we are the customer, and while the insurance company, through their agent, have been happily cashing our premium cheques, they don't want to part with the goodies. We have to swing our attitude 180 degrees and we are slow to react – the battle continues.

We rent an Atco-type trailer, install a temporary electrical service panel, put in a little stock and continue to pump gas, but we lose money every day. We only have Candy and Donna now, and we wouldn't need them but for the fact that we are lawyered up and often on the road.

One nice, clear, sunny day we are heading to Winnipeg to see our lawyer. The straightest shot to Winnipeg runs 40 miles through Minnesota to the Manitoba border, skirting the south shore of Lake of the Woods.

We motor along through Minnesota (speed limit 55 mph.) The cruise control has quit working a couple of months ago and we drive along discussing our problems, as we have been doing daily for a year now. I get more and more agitated and my foot gets heavy. Oops – I'm doing 75! I back off to 55, the discussion continues, and I'm back doing 75 mph. On the third cycle a cruiser comes out of nowhere and his lights are flashing! Uh oh, I pull over and a young state trooper comes to the door. He calls me sir and tells me the air patrol has clocked me at 75. Holy cow! I didn't know they had an eye in the sky. I figure my goose is cooked but I tell him my side of it anyway. I figure it's worth a shot.

The trooper listens to my tale of woe and invites me to come back to the cruiser: he is going to talk to the pilot.

I join him in the Crown Vic and there is barely enough room for the two of us. The space in the middle is loaded with at least three rifles and who knows what else. I am impressed – this guy can quell a riot single-handedly. He gets the pilot on the radio and once again I am impressed. The pilot has everything down pat almost to the V.I.N. hidden at the base of my

214

windshield. I'm pretty sure he could see where I cut myself shaving. He must have eyes like a hawk.

The pilot confirms my story – up to 75 then back down to 55 – three times, he says. The trooper is almost apologetic as he writes the ticket. After all, I was way above the limit. He gives me the absolute minimum, 56 miles per hour. The record will die in Minnesota, I won't lose any points in Ontario, and our insurance rate won't skyrocket. I've hit the jackpot again! As I exit the cruiser he wishes me luck. Jeez, I love a good cop!

Two days later Ken Badiuk fixes the cruise control. I won't tempt fate again.

"Good Cop" Story
A retired cop and good friend lives nearby, and he has a story or ten. Earlier in his career he is pulling a night shift in a sleepy little town. He's parked near an intersection, it's 3 am and nothing is moving – nothing. A guy comes along and blows the stop sign completely, doesn't even slow down! On go the flashing lights and my friend pulls the guy over. "Didn't you see that stop sign?" asks my buddy.

"Sure I saw the sign," says the guy, "But I sure as hell didn't see you!"

The guy got a warning. My friend had to get back to his cruiser, where he could split a gut in private. Good cops are born, not made.

Same Cop, Different Story
A motorist had wiped a section of guard rail and my friend attended: turns out the guy is a deaf mute and notes must be exchanged. My friend writes out "What happened?"

The mute puts his hands together and lays them on his cheek, indicating he fell asleep.

The cop writes, "Falling asleep – dangerous driving, loss of points, possible loss of licence, insurance rates sure to go up. Swerving to miss dog – excusable."

Deaf guy takes the pad, writes "Swerved to miss BIG dog."

My friend felt that since no damage was done other than to some Highways Department property, why should he nail the guy? He already had enough problems in his life.

1996 is not a good year at all. We are still at an impasse; we are losing the business and I'm losing friends. N.B.U. Jack is gone and I'm spending nights at Sam's. He is afraid he will die alone, and his family is tending to his ex who is also ill. I don't mind. Sam has a ton of stories to tell and I wish I could remember all of them.

Table Scraps: as told to me by Sam Duggan.

Firearms Training
Sam and his wife, local prospector Nolan Cox, a geologist from Toronto, and the local bank manager were having a pleasant evening in the bar at Beardmore. Also in the bar at the time was a big native guy from a community on the shores of Lake Nipigon. He was pretty drunk and was hassling the geologist. Sam said he knew the guy, and that he was usually easy going, but for some reason he didn't like the geologist, and wouldn't get off his back. Sam even had a chat with the guy, but to no avail.

The party was breaking up and Sam and his wife walked home, about two blocks away. As they got to the house his wife said, "Is that thunder?" Sam said it wasn't even cloudy.

They no sooner had taken off their coats when someone banged on the door. There was the bank manager, out of breath and very upset. "Nolan just shot the Indian!"

Sam and the bank manager beat it back to the bar. There was Nolan, the geologist, a gathering of people, and a body (??) on the sidewalk.

What happened was, as Nolan and the geologist were walking back to Nolan's truck they were followed by the belligerent guy yapping at them. Nolan was fed up! He reached into his truck, took his shotgun from the rack behind the seat and trained it on the guy. "Shut up or I'll drill you!"

The guy kept yapping and Nolan pulled the trigger.

The guy gasped, "You shot me!" and dropped.

When Sam got there, someone opened the guy's vest and said, "I don't see any blood."

Earlier that day Nolan had taken his young Labrador retriever out and was conditioning him to the sound of gunfire. To do so, he had taken the pellets from a number of shells, and the gun was loaded with blanks. When Nolan pulled the trigger, the wad hit the fellow a good whack, he thought he was shot and he fainted. He soon came around, realized he was OK, (a little sore, no doubt) and staggered off home. Sam told me (perhaps it's true) that the guy never drank again.

I don't know how Nolan's retriever turned out, but I think you will agree that the training was successful.

We've Struck Wienerite!

The day after the training session the geologist and Nolan were planning to visit the drill site on Nolan's gold claims. Nolan was quite enthused about the results so far, and had high expectations of talking the geologist and his company into a decent option agreement. Of course, there had been trenching, sampling and other work done, but the drill core would tell the story. One box in particular looked pretty darned good!

Let me explain for the uninitiated: Drill core is retrieved from the drill hole via a core barrel and the core is placed in boxes in five-foot sections, in the same order that it came out of the ground. The boxes are eight inches wide and have round grooves for the core to lie in. For transport, an identical box is fastened upside down on top of the cores and held in place with two wraps of wire, making a tubular pocket that holds everything in place. (Think of a giant egg carton with long grooves instead of pockets.) This was "A" core, 1 1/2 inches in diameter, so a box would hold four lengths, or 20 feet per box.

The drillers had the core ready to move, but Sam knew Nolan. He wouldn't be able to resist opening up the best box to show the geologist.

Sam and another friend Dave had beat it out to the drill, hid their vehicle and replaced the box in question with another marked with the same footage information. In this box they put 20 feet of hot dogs they had brought, wrapped the box back up as usual, and hid in the bush.

Nolan and his man showed up on cue and looked at some of the core. (Nolan was building up to the big scene.) He pulled out his prize box, undid the lid and removed it with a flourish, crying, "Feast your eyes on this!" while watching the geologist to revel in the expected reaction.

He got a reaction, all right! The geologist just stared. Nolan turned back to the core box, saw 20 feet of tube ssteaks, and Sam said it was priceless.

Nolan's eyes got big and he stood there like a fish out of water, his mouth opening and closing, but no sound coming out. When words did come out, they were unprintable.

Sam and Dave came out of the trees, laughing like crazy.

"You've struck Wienerite!" they hollered.

Workplace Health and Safety

In the early eighties Sam is living in Fort Frances. He is sinking a shaft on the Nickel Lake copper deposit, (remember Chick and Leo) and although the company will soon run out of money, work is still underway.

One evening there is a knock on Sam's door. He opens it to find an officious looking little gent who says he wants to talk to Sam, so Sam invites him in and offers him a drink.

He identifies himself as a Safety Assessment Officer with the Department of Labour, and tells Sam that he will visit the construction site tomorrow. Sam is a bit mystified – these visits are never pre-advertised – it sort of negates the surprise factor, but Sam is interested in what the guy has to say.

After 15 minutes of beating around the bushes the true purpose for the evening's visit emerges. If Sam will pony up X amount of cash, tomorrow's assessment will be tickety-boo. Failure to do so will result in many infractions being found – in other words, "Pay up or I'll shut you down."

Big mistake! Sam is no spring chicken anymore, but he is mad! He picks the guy up by his collar and the ass of his pants and tosses him out into the front yard.

The guy turned out to be a pretty good assessor of his own safety. Sam never saw him again.

Hans vs the Investment Club (Sam again)

I met Hans in 1969. He was a Red Lake old timer, a friendly guy, always circulating the bar scene, meeting new people, picking up tidbits of the Uchi Lake rush, and always willing to share his large store of memories and info of everything that glitters. Hans had been here since the hectic thirties, had made more than one find, had seen more than one head frame rise on properties he has dealt, and like many of his compatriots, had played the stock market over the years with mixed success. Sam has a couple of Hans stories for me.

In the fifties there was an informal investment club in Red Lake: no big deal, just some guys pooling a bit of cash and more interested in the social aspect rather than hard-nosed business, if you catch my drift. The meetings were informal, usually held in the back room of any number of business establishments. The dues where B.Y.O.B, and Robert's Rules were never considered. The floor belonged to whoever could talk the loudest and longest and Hans, who could tip a jar with best of them, usually had the floor. The problem is, Hans is sort of an I've been there (before you) and done that (better than you) type of guy when he is in his cups, which tends to piss the others off, and tonight he has been more obstreperous than usual.

So Hans over-indulges and decides to have a nap on the storeroom floor. The rest of the boys (not a sober one in the bunch) decide to play a harmless trick on Sleeping Beauty.

Tonight's meeting hall is a storeroom behind a hardware store, so it's easy to rustle up some small nails and a hammer. Ever so carefully and neatly they nail Hans' clothing to the floor, taking care to fold the cloth tightly beneath him, and when the last nail is pounded in Hans is pinned from ankles to arm pits, and down to his wrists.

The last bottle is empty, but the Bucket of Blood is still open and the investment club decides on a change of venue. Hans is left to sleep it off.

Now, this is the time of wood heat, and while the rest of the guys are making millions in the Bucket, the wood stove in the hardware store dies down and Hans wakes up. He can wriggle his head hands and feet, but he is sure he has had a stroke or some such. He is paralysed, he is alone, and he is getting cold. Hans figures he has been left to die alone.

The last stock has been traded, and last call arrives before someone remembers poor Hans. They beat it back to the hardware and find him half frozen, with tears running down his cheeks. It's not easy to pull nails out of the hardwood floor, and they have to cut him free. Someone finds a blanket and they take Hans home to bed. He has been scared sober.

Another investment club meeting: It's a warm summer evening, but the gathering is rather cool. A few meetings ago the club, at Hans' urging, has invested heavily in a stock which has tanked, and tonight they feel that retribution is nigh. Various methods of punishment are discussed, and as the bottle levels get lower the members' pissed-offedness increases. The jury comes to a unanimous decision – this is a hanging offense. Hans of course doesn't agree, but he is outnumbered.

He is marched out on Main Street with his hands tied behind his back. Someone has a rope (in any frontier town worth its salt a rope is always available.) A noose goes around Hans' neck, the rope is thrown over a telephone pole cross arm and tied to the back bumper of a '52 Ford. Hans is pleading for mercy and the cavalry arrives in the nick of time – someone has called the cops. Hans gets a last-minute reprieve and the club adjourns to the Bucket.

Would they have hung poor Hans? Well, maybe yes, maybe no – we'll never be sure.

Our insurance battle goes on – but every hurricane has an eye, and one weekend stands out bright and clear.

It's a Saturday morning in mid-September and I am sitting in the Atco, waiting for the occasional customer. Dwayne Phulak goes by in his beat-up Dodge pickup towing his modified on the hauler. I wave to him and wonder where he's going – our local track shuts down on Labour Day. A few minutes later, much to my surprise, Dwayne comes back and pulls into the parking lot. What's up?

Dwayne says we have to come to Winnipeg! It's Enduro Weekend at the Winnipeg Speedway, a one-mile dirt track which my wife and I had often enjoyed in the early 80's. Twenty-lap qualifiers start at noon today. Sunday they will run three classes: 50 miles for the hobbies, 100 miles for the streets, and 200 miles for the modifieds. Seven of our local modifieds are on their way, and the track expects 100 cars for the 200 lap feature.

I tell Dwayne I can't do it, someone has to run the store. What I don't tell him is that we are dam near broke. I watch him head west and I feel really, really bad.

Let me tell you about Dwayne: When I worked at the mill in '66 -'68, I always put up the hay for Dad, and in 1967 I hired Dwayne for a couple of days to help haul bales. He was 14 years old then, but he was a strong kid and could throw a bale as far as any grown man. He was already a car guy and when I tried to pay him he refused to take my money. Instead he took my retired 54 Plymouth home and drove it around in circles until it blew up. In 1968 he put his first stock car on the track, and by 1996 he has been racing for 29 years, and has been top man more than once.

But Dwayne is a low budget racer. Although a few local guys with deeper pockets have enclosed haulers, most pull their cars around on 2-axle open trailers, with the nose of the car nestled under a tire rack holding 6 or 8 spare tires. Dwayne has two spares.

So I sit in the Atco, pump a little gas and think and sulk. I pull out my last surviving credit card and study on it a bit. It's pretty much maxed out, but I think there might be room for a couple of hundred bucks.

What the heck – I call Donna – can she cover the pumps today? Yes she can. Cool - tomorrow is Sunday and we are closed on Sunday nowadays. I call my wife and tell her to pack an overnight bag – we are going to chase some dirt. Within an hour we are following Dwayne west.

We hit the track at about 1 pm, and it's packed! In the infield haulers are lined up cheek by jowl three rows deep. This will be a good show for sure!

The hobby and street qualifiers are done by the time we get there, now the modifieds start. They will run 5x20 lappers of 20 cars each to determine starting position for tomorrow's race.

It's a long afternoon before the lineup is set. Some of our guys do pretty well. One is in the top twenty, others a bit further back. Dwayne doesn't do so good. He's not on the tail end, but we will need binoculars to pick him out. The pole-sitter is the 8-Ball, from Fargo, North Dakota. He is the current Wissota champion, and is the star of the show. This is no low budget racer, let me tell you. His hauler truck would not be out of place at Daytona.

My wife and I get a room at the Comfort Inn and that evening we join Dwayne and his crew chief for a drink. (Crew of one – Ben Brown has pitted for Dwayne for years, and is one sharp cookie). The wife and I don't have much to say – we listen and learn.

The issue is rubber: there are only three decent tires left on the car now. One is worn out, and the other three less so, and one of their spares is equally sad. They have one new tire which they have saved for tomorrow, (this explains the poor qualifying effort), the question being: Where will the new tire go?

Dwayne wants to put it on the right rear. Ben disagrees. He thinks it should be on the left, and he tells Dwayne why. If they put the new tire on the right side it will probably wear down a lot before the end of the race. On the inside it will barely touch the track on the corners, and when the car settles down on the straights Dwayne will have a decent dig at the track surface. The decision is made – left rear it is.

Sunday morning, and it's going to be a good day at the races. It's nice and warm, and a thin overcast will protect the track from the hot sun. Dust won't be much of a problem.

The wife and I arrive early to make sure we get our usual spot at the end of the front straight at #1 corner.

The hobbies and streets run their races, and they are dandy. No matter what the class, if the cars are competitive, the race is enjoyable. The street stocks are definitely a crowd pleaser. Halfway through the race a mean looking old Impala big block starts to smoke, and although he doesn't slow down too much, the smoke gets thicker and blacker. He looks like he's burning coal. By lap 98 the smoke clears somewhat, but now sparks are flying, and when he passes the grandstand the smell of hot metal lingers in the air. He makes it to the finish line, and when he hits the chequered flag that sucker flat explodes! A mushroom cloud rises over the Winnipeg speedway and the crowd goes goofy! It doesn't matter where he finished the race as far as we're concerned – he's the winner! The poor old Impala goes back on the hauler. The guy has to hope his wife will buy him a motor for Xmas.

They water down the track and a hundred modifieds line up and start their pace laps. All the colours of the rainbow are streaming around the track. Some are solid blues, yellows, and what have you. Others have fancy graphics, and all are nice and shiny, but not Dwayne.

One rear quarter panel is black, the other sort of a maroon color. The hood and roof are white and the rear deck doesn't match anything. His sponsor's names are faded or half obliterated by tire rubs. The recently-replaced right side door panel is unpainted, and his car number is done with good old duct tape. Maybe there's a method in his madness. Like a beater in a Lamborghini parking lot, he has lots of space around him and track real estate is pretty scarce. When they drop the green flag the tail end cars are in the middle of the back straightaway.

There no way to do justice to the sound of 800 methanol-gulping, ground-pounding cylinders, except to say WOW! Everyone is on their feet and the grandstand is vibrating. We all settle down after lap one, but 3000 right feet are pressed to the floorboards. We're racing baby!

8-Ball takes the lead in turn one, and it looks like he's going to hold it. By lap 20 the front runners are strung out pretty good. Not in the middle of the pack though. Inevitable tangles happen and some cars are hauled to the infield, and the restarts close up the field: 8-Ball doesn't get much breathing room.

Attrition takes its toll – a blown engine here, a transmission failure there, and the field gets a bit smaller. The announcer is in love with 8-Ball who gets most of the press, but to be fair, I have to say he treats some of our guys pretty good. Our local racers come to the speedway quite often, and the announcer knows that our little track has some hard chargers.

We've been cheering for all our boys while keeping a sharp eye on Dwayne, and he's been picking up a spot or two every lap. There is a mandatory yellow at the 100-mile mark, and Dwayne pulls into the pits in the top 30. All the drivers put on fresh tires and take on fuel. It's fuel only for Dwayne. He has no fresh tires.

The race restarts and our best qualifier is in the top 10 and coming on strong. Disaster! He blows his motor and retires. Only three of our guys are still running, and two of them are far back in the pack. After 150 laps I think Dwayne is fifteenth, it's hard to tell now with so many cars having been lapped.

Finally, the announcer starts to figure out that something is going on here: Who is this masked man, anyhow?

At first it's obvious that he has only the car number, but I guess someone looks at page five of the start position sheets and comes up with Dwayne's name, but now he can't pronounce Dwayne's last name. The home fan contingent puts up with it for a few laps and then a lady trots up to the booth and straightens out Mr. Tongue-tied. Now Dwayne's last name is being dropped as often as 8-Ball's.

Dwayne is still cherry picking out there. The lapped cars are easy pickings, and even front-runners pose no problem. Maybe they all put their good tires on the outside rear.

At mile 185 Dwayne is running seventh and we are on the edge of our seats, as is everyone else. I'm pretty sure the whole crowd is on Dwayne's side by now. He hits the end of the front straight, and right in front of us his motor dies. Not a burp or a miss, just flat dies, and WWII could be at its peak out there, but to me the silence is deafening. My heart skips a beat, but wait! The car coasts silently for 50 feet, gives off a puff of blue smoke (Dwayne's engines are always old enough to smoke,) fires up, and Dwayne is back in business.

(Here's the deal as I later learn. Dwayne keeps a box-end wrench beside him in the car. By reaching back to his right he can adjust the track bar during the race. Is it legal? Maybe yes, maybe no – it's certainly effective, but when Dwayne reaches over this time, he hits the toggle switch and kills the engine. When the motor dies, Dwayne dam near croaks also, but when he sees what he has done he simply hits the switch and goes back racing.)

On lap 193 Dwayne picks off the second-place car and starts to reel in the 8-Ball, and on lap 195, he's tucked in under 8-Ball's rear end! On lap 197 he sets up 8-Ball coming in to turn three, and comes out of turn four in the lead!

Well, you should have heard the crowd! We are all on our feet hollering! It could have been Dale Jr. out there! But 8-Ball gets his nose in on the one-two corner and takes back the lead. For the next three laps we all wait for Dwayne to take another shot at it, but it doesn't happen. The chequered flag is out and 8-Ball crosses the finish line first, with Dwayne in second.

Now, there's two ways of looking at this. They say you don't have to lead every lap, just the last one. Well, 8-Ball led 199 laps – the last one for sure. But I look at it this way: 8-Ball ran 200 miles with little to worry about except some lapped stragglers. Dwayne ran 200.5 miles, avoiding wrecks, passing 78 better qualifiers (79 if you count lap 197,) and finished on the same tires he started the race with. Like Cale Yarbrough, he could take a second rate car to a higher level. 8-Ball might have won the race, but I'll always think that Dwayne was the top driver that day.

A couple of weeks later I run into Dwayne and get his side of the story. He tells me that he was as surprised as anyone to find himself on 8-Ball's back bumper, but the most startled guy on the track must have been 8-Ball when Dwayne pulled him on lap 197. Dwayne gave the track back to 8-Ball because he didn't quite trust him. For one thing, he's never raced the guy before, and he's not sure of 8-Ball's modus operandi. Dwayne figures 8-Ball's tires are going away and he can be had, but what if he cuts down on Dwayne and they wreck?

There's also the prize money to consider – 800 bucks for first, 500 for second. The money means nothing to 8-Ball, only his pride is at stake; Dwayne has a thirsty old Dodge to feed for the trip home.

Can you imagine? Tooling around wide open on a one-mile dirt track with the Wissota champion lined up in your gunsights while juggling all these options? It boggles my mind for sure, and I'm not behind the wheel.

To add to Dwayne's problems is the fact that he's not too sure how many laps are left. They have a girl on a chalkboard near turn one, but she's a fan also, and is not always up to speed. By the time Dwayne decides to take another shot at it the flag drops and the race is over.

No big deal though, the room and the trip are covered. The car goes home to the shop to wait for next year. Maybe there will be enough left over to put a new tire on Number 21.

Now, the wife and I have chased our share of dirt, we've seen modifieds at the State Fairgrounds in Minot, North Dakota, made more than one trip to the late model invitationals at Cedar Lake, Wisconsin, and have watched Tony Stewart's Prelude to the Dream late models on the tube, but in my mind on that Sunday afternoon in September, we were privileged to watch the best race of all times!

I can picture every scene in my mind, every sound, and the smell of methanol is in the air, and I still get all tingly feeling. Twenty-odd years have passed, but it could have been yesterday. Pretty cool, I'm thinking.

Dwayne is retired and has sold his racer, but he still turns left some Saturday nights. He mentors a young buck or two, helping to set up their cars. He'll run the odd race to get the feel of the car, but of course tries to avoid tangles – he doesn't want to mess up their spiffy paint jobs. But dang it all, I wish he would go retro to his coat of many colors. I miss it.

While I'm on the subject I might as well tell you about the second best race I ever watched – a five lapper.

Tonight our local track is running a celebrity race following the modified feature. The crowd stays put - we want to watch this.

There are only 6 or 7 cars involved, all modifieds. I can't recall three of the entrants, but they are an eclectic bunch for sure. A local O.P.P. officer, Dave Lee, is running, and there are two prominent politicians in the field. One is our federal Member of Parliament, Robert Nault, who hails from Kenora and the other is Howie Hampton, our provincial M.L.A. Howie is a third-generation valley boy – everyone knows Howie. One of the drivers is a ringer. His name is unannounced, but everyone in the grandstand knows who he is by his car number. He is a dandy racer, but sort of a free spirit, if you catch my drift. He's been a bad boy, got caught running a roller cam and has been barred from the Wissota ranks. He has brought his car for one reason tonight – he wants to chase a cop for a change.

So, fire suits are borrowed, flag colors are explained, seven cars take their pace lap, the green flag flies, and Howie drives straight off the number one corner, over the embankment, and into a mud hole.

The wrecker puts the hook on Howie, drags him to dry land and he does a couple of laps to shake off some mud. The green flag drops again. Howie puts his foot in it, and hits the mud hole. The crowd goes goofy! A guy sitting ahead of me plays liniment league hockey with Howie, and he tells me things are not much different on the ice. Only the end boards stop Howie.

The green flag flies again, and once again Howie does his exit strategy. This time the flagman has a chat with him. Howie is a socialist. Maybe he should follow his political instincts and turn left once in a while. Howie agrees – he is clear on the concept now.

The rest of the race is good clean fun. Free Spirit and the cop play tag and have a great time. The Kenora M.P. although politically liberal, is definitely conservative on the track. He doesn't want to tear up his borrowed car. I don't even remember who won the race. No matter, we all go home feeling we've gotten our money's worth tonight.

Our battle is lost. We shut down Christmas Eve and clean out the Atco before the New Year. In January we declare bankruptcy. It is absolutely unavoidable. We are penniless; having sunk our last cent, plus other people's money into a fruitless fight. A recalcitrant insurance company, unethical insurance agency and incompetent and unscrupulous lawyers have tapped us out. We can't even look our friends and supporters in the eye, so we cash out our one remaining R.R.S.P. and head west.

Chapter XIII   How the West was Formed (among other things.)

For the next nine years we boot around trying different ways to make a living. This, that and the other thing, while we try to get our heads in gear, but with mixed success. However, one effort to establish ourselves as functioning members of society is worth a line or two. The job itself is hardly worth mentioning – once again it's the people we meet and hear about. Some are good folks, others not so much.

We take a crack at managing a lodge near a native community on the east shore of Lake Winnipeg. It is what is called a remote community, serviced by freighter and barge in the summer months and by a winter road. A busy airport connects us to the outside world year round – a handy thing during freeze-up and break-up.

The lodge is a huge building – 40 x 80 with a 20 x 30-foot addition containing kitchen facilities. On the main floor there is a large sofa-filled TV room, a dining room and living quarters. The second floor has fifteen guest rooms – doubles and singles.

A few years before a bar had been added, and it can only be described as a hell-hole. The Bucket of Blood back in Red Lake in the '60s was Sunday School compared to this joint. It will be the main reason we will give up on the place six months later.

We are successful during our tenure period. The palace actually shows a profit for the owners, mostly due to my wife's work ethic. She runs the kitchen and cleans the rooms virtually singlehandedly.

We have one ally – Wesley. He is a community elder and drives the community water truck. Wesley lives alone in a neat bungalow, delivers us water twice a week and visits us for evening coffee. He is a storehouse of historical memories, has many stories to tell, and great stories they are.

Sidebar: As co-authors, it has been our writing style to feature slightly amusing and interesting tidbits as Table Scraps, but the term would do injustice to Wesley's tales. Most of them are legends as passed on by elders from generation to generation.

I hope you can picture the storytelling as I did. We are not in an up-to-date dining room. We are gathered around a warm stove in a trapline cabin where three generations are listening to Grandpa before snuggling into bed. Outside there is a soft snow falling – inside the stove crackles as the legends flow.

Wolverine and Skunk

The time was long ago and the land was different then. Lake Winnipeg was there, having shrunk to its present size following the disappearance of the huge Lake Agassiz, but westward to the Pacific, the land was a mixture of outcroppings and stunted trees – hardly worth looking at and populated by big scary creatures.

The forest on the east side of the lake was healthy and mature, and the animals were likewise. Most had their ways, habits, and their place in nature figured out, but others were still adjusting. Skunk, for instance, was an aggressive hunter, and Wolverine, at that time was a team player.

One day in early spring Wolverine was passing a beaver dam when he saw Skunk digging into Beaver's house. Wolverine knew what Skunk was up to – he was after Beaver's children. Wolverine liked Beaver and his family and tried to reason with Skunk. Skunk ignored him and continued to dig into the lodge.

Now Wolverine also knew about Skunk's self-protection capabilities. If he chomped Skunk's ass so Skunk couldn't piss, maybe he would be able to pull Skunk out of the lodge and toss him away. So he did, and Skunk grabbed ahold of some branches in the lodge. Wolverine pulled, Skunk held on, and they were deadlocked. Now Wolverine had a problem – he didn't dare let go!

Lynx happened to pass by, stopped, and Wolverine asked for help. Lazy Lynx said that he couldn't help much, so Wolverine (talking out of the side of his clenched teeth) told Lynx to round up some other animals – Lynx left to do so.

It was a warm, sunny afternoon. Lynx, being a member of the Cat Clan, could not pass up a chance to stop on a nice south-facing slope – Lynx fell asleep.

Wolverine's jaws were aching. He let go and ducked. Skunk pissed!!

And did he ever piss – scouring the land to the west. Rocks flew westward and piled up high before they reached the ocean. All remaining rock was pulverized into soil and gravel – flat as a pancake. The big scary creatures were all buried – never to be seen again.

By now the wife and I are laughing like crazy and really enjoying this. Wesley is unperturbed. He knows we are doubters, but he points out the prairie alkali lakes. White folks call it alkali – the People know it's Skunk piss.

There was more to the fable. Skunk switched his diet to small rodents and roots and was an outcast in the animal world.

As for Wolverine, he became a bitter loner – one compensation for Wolverine was that when the piss passed over his back, it turned his back hair silver and left it permanently super-heated and impervious to frost. Wolverine could now sleep alone in the winter – unharmed by the coldest weather.

Nanabush

Wesley will now tell us a Nanabush story, but first we need some background re: the Spirit World. As Christians, we find it a bit difficult to grasp the concept. After our lesson is over we realize that deep inside Wesley is of the People, but takes care to keep his beliefs to himself unless he trusts you. We feel honoured.

Sidebar: What Wesley does not realize is that he is actually more "Christian" than most. He is never critical of our take on life, and only mildly critical of his own kind as they struggle with the modern world. He may himself have been a victim of the residential school fiasco, but never brings it up. Perhaps his family escaped to the trapline – we will only be able to guess that part of his back story.

So Wesley tries to explain Nanabush to us and to do so he has to bring up WENDIGO and it is clear that this makes him uncomfortable. Wendigo is the Supreme Spirit and a harsh dude. The People are never to speak his name and Wesley is taking a big risk (as are we,) and anyone who has seen Wendigo's face will never describe it. They will see him only once, when Wendigo ushers them into the Spirit World. (And my spine is prickling now, as it did that evening.)

Nanabush, however, is a benign Spirit/Man. He is the go-between go-to guy, usually kind, and is the people's intermediary. He can temper Wendigo's temper and facilitate the transition to the Spirit World. Because he wants to understand humans, he often tries on their moccasins.

Nanabush wants to experience blindness, as some of his People are afflicted. To do so he puts on a blindfold and goes for a walk in the bush, asking the trees to warn him should he be heading for trouble. Evergreen and Poplar guide him carefully, telling him if there is a deadfall or pothole ahead. Then he enters a copse of birch and tag alders near a beaver dam. Birch and Tag Alder are mischievous. They want to give Nanabush a dunking, so they tell him he's doing fine.

He is getting close to a beaver dam and Beaver sees what Birch and Tag Alder are up to. Just before Nanabush steps into the water Beaver slaps his tail and Nanabush steps back.

He takes off his blindfold – sees he has been fooled and his human side gets very angry. He borrows a branch from Balsam Fir and gives Birch and Tag Alder a whipping. To this day you can still see the marks left by the needles on their bark.

Beaver deserves a reward. What does Beaver need? Beaver tells Nanabush that his back gets cold in the fall because it rides a little above water when Beaver is swimming. "No problem," says Nanabush, and now Beaver has an extra layer of fat along his back.

What wonderful stories/legends. I wish I could remember more, but Wesley and I both enjoy an occasional tipple, and when the jar gets low, so does my memory bank account.

This is a mixed community, roughly 50% status and 50% métis. Government infrastructure funding is more readily available for the status side. Thus one half of the population has a water treatment plant. The other half has wells and/or the lake to draw from.

I ask Wesley how the lines had been drawn in the past. It seems to me that complexion-wise there is no rhyme nor reason.

He says that when the government decided to status-ize the People it was left up to the parish priest to decide who was a "Real Injun." The priest's criteria was the Sunday collection tray – if you donated you were STATUS.

We notice a strange phenomenon. In the early evening when driving by a dishevelled-looking house, we can see inside a brightly-lit living room, curtains not yet drawn. The interior looks neat, clean and well-furnished, belieing its raggedy exterior. Wesley tells us why.

He says some of the people tend to be jealous. If anyone makes an obvious show of being upwardly-mobile, they stand a chance of arson. They make themselves comfortable inside – outside the grass stays un-mown.

Change Paradise? Put up a Laundromat. (With apologies to Joni Mitchell.)
Wesley tells us that here had once been a huge stone fireplace in the north wall of the lodge. Now there is only a conventional steel stove and Selkirk chimney. How and why was the fireplace replaced? Wesley says it was a do-it-yourself project

To increase cash flow, the owner had put a laundromat including propane-fired dryers in the basement directly under the TV room and entrance foyer. One day a customer complained about a strong propane smell – Owner investigated.

There are two ways to check a leaking propane fitting, one being to spray the fitting with soapy water. The other way is to light a match – Owner used the sure-fire method.

He was in the centre of the explosion and believe it or not, he only singed his eyebrows. The north end of the lodge did not fare as well – the floor lifted two feet and the wall, including the fireplace collapsed outward. There was no fire – when the floor fell back it extinguished the flames. The only injury was to some poor guy who happened to be standing on the floor and he only suffered short-term leg compression.

Wesley also tells us that while they were standing and studying the new north view, a guy climbed up into the stairless and doorless lodge and asked if the bar was open yet.

How to Give Customers an Unforgettable Wilderness Experience.

I noticed some stuff in the storage shed – a beat-up rubber raft, some old camping gear, and paddles. As usual, Wesley has the explanation.

Owner decided to jump into the eco-tourism game. He had a Beaver, a beautiful stretch of water on the Pigeon River and a plan.

He would outfit groups of four, fly them upriver to a wide part of the Pigeon, and they could float/paddle leisurely westward to Lake Winnipeg with an overnight stop or two. Then he would pick them up at the mouth of the Pigeon 40 miles south of the Lodge. It would be a unique, pleasant experience for urban folks. What could go wrong? Nothing – if Owner had not suffered from attention deficit disorder.

The first few trips went well, were enjoyed by all concerned, and word spread - then the inevitable A.D.D. struck!

A group arrived at the mouth of the Pigeon and were <u>not</u> picked up. They sat there for three grubless days before a passing boat saw their distress signal.

Goodwill news reverted to badwill, and the eco-tourism venture was as dead as a pigeon.

A Fairly Normal Night in the Bar

This my tale. I was there, saw this, and they done that.

It's January and commercial fishing is underway on the big lake. An imaginary apartheid line twenty miles south of us divides the lake – whites on the south, natives to the north.

A Gimli group of three stay at the lodge as 20 lake miles by bombardier is just a short jaunt. The head guy is B, his main man and long-time friend is J. They fish by day and party by night.

They leave early and return by 3pm. Our bar opens at 4pm and B and J are first in line.

As always, it starts out quietly – beer and conversation between two old friends. The pace starts to increase and an argument starts. B stands up, pokes J in the nose and both J and his chair end up on the floor. No problems, mate, J picks his chair up, sits down and things return to normal – not.

Another flare-up, and this time J knocks B on his bum. Back to the chairs. Two rounds later I shut them down for a meal break – they are already late for supper.

I'm enjoying this stuff and what's really funny is the sight of my normally rowdy regulars. They want no part of these two boxers and stay away from the ringside seats. They sit as close as possible to the outer walls and avoid eye contact.

B and J sit down in a dining room booth and sort of eat supper, which has been kept hot for them. Do they settle down? Not on your life – they eat a bite, argue, and disappear outside for a punch-up and repeat – three times. I am still enjoying it but my wife puts the hammer down and sends them off to bed.

It's a good thing they are our only guests tonight. Their room is directly above our living quarters and for two or three hours we hear thumps and bumps.

In the morning they come down for their usual early breakfast, badly hung over, but pals again. Neither one has a mark or a bruise. Maybe they have stunt doubles.

I ask them if they had continued the fight upstairs – B doesn't know what I am talking about. J said he was the culprit – his bed wouldn't hold still and he kept falling out on the floor. What a dynamic duo!

We have other guests tonight so when the boys hit the bar, I'm watching them – I may have to step in. Then I get a phone call from Gimli – B's wife wants to talk to him. Before I rustle him up I have a chat with the lady. If tonight is a repeat the crew may have to find alternate accommodation.

The ensuing conversation was short and to the point. B got J out of the bar and for the rest of their stay they were good little boys.

(I later found out that while B was considered the commercial fishing champ in Gimli he also held the local drinking title. His wife (by court order, no less,) handled the money. If B wanted to party, mommy would cut off his allowance.)

Pleasant Interludes

Wesley always had coffee with us at 10am. Out on the bay a group of eight to ten dogs would head south at that time. Wesley said they were off to have lunch??

The dogs were smart. They knew that the nets would be pulled by the time they reached the Grand Banks and they could have their fill of trash fish. One dog was a dachshund cross – all body and short legs. He must have had a good cardio workout on the forty mile round trip.

Every Saturday morning Wesley comes for coffee and we play Radio Bingo.

It is a neat setup – a well-oiled machine. Each community has a rep with a route. Wesley has told the lady that we would like to join the bingo club, so she drops in early to sell us cards. The wife and I take six each – Wesley plays twelve.

So we drink coffee for a while, spending our jackpot, and turn on the radio at 10 o'clock sharp – we don't want to miss the fun,

The station is at Thompson. It's all aboriginal, totally run by natives, broadcasting northern oriented news, country/western/gospel music, and does a great job. On Saturday mornings every ear in the north is tuned to Bingo time. Today the blackout jackpot is carried over.

We called Wesley a couple of years later, He told us he had won 30 grand on the blackout. Knowing Wesley, I'm sure he shared with his two married daughters. Perhaps he upgraded his old Chevy S-10 pickup to an S-11. Like houses, fancy vehicles tended to invite vandalism.

The next time we called we were unable to track him down. We assume Wesley had an appointment with Wendigo.

We learn some interesting stuff about other folks along with facts of life pertaining to the People.

Table Scrap

Owner has a 27-foot steel boat suitable for freighting. He has a local captain – a young fellow who we will call Frankie. It's a two-hour trip to Matheson Island and Frankie can navigate Lake Winnipeg blindfolded. The six-cylinder boat engine runs on diesel – Frankie runs on gas.

A witness saw the whole deal. Frankie, out of gas, smoothly rounded the point into Matheson Harbour and just as smoothly ran the boat up onto a rock reef – high and completely dry. He tried to back off the rock with the prop only grabbing air and fried the six-cylinder motor – a $6000 dollar rebuild.

A month after our arrival I go to the airport to pick up 20 cases of beer. (With no boat the beer is now airfreighted.) A young passenger helps me load the beer and asks for a ride home. Why not – he has been kind enough to help me.

He introduces himself on the way. He says he is the ship's captain.

"They still owe me $35 for that trip," says Frankie.

Table Scrap

I meet Goosehead in the bar one night (I seldom work the bar. The drunks see me as a challenge and I don't like to fight.) Goosehead introduces himself – he is a happy-go-lucky fun-loving young man. One of our brands is Moosehead, a New Brunswick beer. "Here's a Moosehead for Goosehead," says I, and he laughs like crazy, almost falling off his stool.

Goosehead was absent for a couple of months that winter. Then one day he comes back into the bar with a huge bandage on one hand and tells me the story.

He was in a bar on North Main Street in Winnipeg. A guy was beating on a woman and Goosehead stepped in. The guy was not pleased with having his fun interrupted, produced a machete, and Goosehead had fifty stitches on the mend.

The guy bolted and the cops came. They took Goosehead to emergency, and after repairs were made they charged Goosehead with assault! Pretty harsh.

Goosehead is <u>not</u> upset as he tells the tale. He says that after all, they had to charge somebody. Goosehead is a realist.

I tell Wesley about Goosehead. Wesley says he is already a local legend. Here's why'

Table Scrap

It's 200 air miles to Winnipeg and a ticket is $200. When the winter road is in one can bum a ride – otherwise one is reserve-bound.

So it's summertime and Goosehead is going to walk – 100 miles down the east shore and over another 100 with little chance of a ride. He packs a lunch – two 2-litre Pepsi bottles of gas!

He travels at night, sleeping in the day when bugs don't bother him, and he has two hefty rivers to swim.

It takes him four days to get to Winnipeg with only his lunch to sustain him. The People know that Goosehead is an environmentalist Poster Boy – 50 miles per litre.

Our managership did not end amicably, nor did it peter out – more like a bang. The winter road is still firm in late March but will soon become impassable, so Owner sends in another 20 cases of beer in the company van – a baffed-out vintage Caravan. He picks a good driver to bring it in – Fast Frankie.

They show up a t 5:30pm – Frankie, Owen and 18 1/2 cases of beer – shrinkage, you might say.

They help me unload the stuff but the last case sitting on the back step does not want to come into the lodge. Frankie has a grip on one end and me on the other. He says it is payment for services rendered. I point out that they have already imbibed their paycheck. Frankie gives one last tug and wins the battle – but not the war.

I've had it, and I launch off the step like Batman! There I am, sixty years old and rolling around in the spring mud with a twenty-three-year-old. Owen has a stick and he is saying that they can kill this old white guy, but Owen is not a brave young Brave. My wife is standing by with a steel bar we use to barricade the door at night. Owen is keeping a close eye on my wife – he knows that an irate woman in full war paint is not to be trifled with. The cavalry arrives in a bombardier. He is a friend of Owner and saw the kerfuffle as he was motoring by. Frankie and Owen simply melt into the bush. I knock off a few hunks of mud and go to see our knight in shining Auto-Nèige armour. Do I thank him? I chew the poor bugger out for driving over our septic field. I just can't help it – I'm still wired.

I resign by phone verbally – very verbally! The next day my wife flies out, rents a Ryder and picks up our personal possessions from storage. The day after I meet her at the airbase on Matheson Island, having flown out in a Cessna 206 with a dog, two cats and our suitcases.

I would never say that the lodge deal was a poor job, but it sure was interesting.

Incidentally – At the airbase the manager asked me if I had been paid in full by Owner. I told him "Dam straight, Last thing I did was to pay ourselves cash and I left the receipt on the office desk."

He laughed and congratulated me, saying that we were the first of a long line of managers to leave fully paid.

230

Chapter XIV: Back to the Bush Again (Sort of)

So it came to pass that in 2005 my wife and I are living near Stratton Ontario. With time on my hands and the news that the Richardson Twp. gold play is heating up, I decide to hit the bush again. Not on the picket line though – this time my grid will be well-gravelled roads – about a forty-mile loop.

I spot a claim post and stop to investigate. I have to jump a bit of a ditch and I almost do a face-plant in the blueberries. The return trip is even more pitiful to watch – I'll stick to gravel from now on.

Things look pretty active compared to ten years ago when we used to do the Nuinsco tour with Sam Duggan and/or Jack Hodge, Mel Jack and John Ross Anderson. There are many drills working and the little coreshack we had visited in '93 – '96 has grown.

There seems to be a competitor drilling a block of ground in the centre of activity. Back in Stratton I ask Robin McQ. if he knows the company and he tells me it is Weststar. I google and find they have a small block of claims on the Split Rock road, but they seem to be on the fringe of the main play. I figure they are keeping their activity in Richardson Twp under wraps for the time being, so I throw caution to the winds and call my son-in-law in Eastern Ontario. On my advice he buys a chunk of Weststar. (It's painless to take a risk with someone else's money.)

On my next tour I see a sign on the road where I assumed Weststar was drilling. The sign says "BAYFIELD VENTURES" – oops!

It's time for damage control. Back at the computer I find that Weststar is involved in a silver play in Western Mexico and I don't speak Spanish. My son-in-law is in Los Angeles for a couple of weeks and I can't reach him – I chew a fingernail or two.

I finally contact my son-in-law and by this time Weststar has been climbing on the stock market and he makes a grand or two or three. We had jumped into a pile of horse manure and have come out smelling like roses.

Thanks, Robin – I owe you a cup of coffee.

I've already filled you in on how I discovered, thanks to Google, that the Bayfield Pres was none other than the son of Carl (who I had rescued at Savant Lake in '71.)

I drop in at Bayfield's core shack, meet Bob Marvin, their exploration geologist and begin a friendship which has lasted to the present day.

As an honorary Old-Fart-Quasi-Prospector I become an interested bystander and am often honoured to be trusted with some privileged information. Not "inside information," mind you – Bob is always careful about that stuff, but I do get some insight into the rough-an-tumble trials and tribulations of junior resource company skirmishes.

Here's a story: To protect the guilty "I'll use a pseudonym – some of you may be able to fill in the blanks.

Big Mouth, for some vague reason, is mad at a group who are often involved in raising money for junior companies and had done so for Bayfield.

Big Mouth has a pal, a "reporter" who works for a scandal rag – a once-respected weekly mining news publication. This "reporter" does a number on Bayfield, balancing on the tightrope of libel. (This I know. I heard half of the telephone interview and Bob Marvin's side was <u>not</u> reported.)

Bayfield stock took a hit and never really recovered.

Subsequent news releases of additional gold reserves would bump the shares a nickle, followed by a ten cent drop. With their credibility damaged beyond repair and with the general downturn in the venture market, the inevitable take-out of Bayfield was relatively undervalued.

There was no collusion on the part of the purchaser (our present developer/operator of the new mine.) It was business, and business is business.

Fast-forward to a recent mining symposium in Reno, Nevada. Bob Marvin attends, and so does Big Mouth. Bob has never met the man and he asks someone, "Where is he sitting?"

He goes to the table, and asks Big Mouth to stand up, which he does. (No doubt expecting a poke in the nose.)

Bob merely tells him a thing or two. He says everyone has a job to do – he just wishes Big Mouth had done his own job and let Bob do his.

I feel much better now, and I am sure both Bob and I feel better than Big Mouth.

I am finding it hard to get out of this darned book. It is about stuff happening, and stuff is still happening, so one more time I'll release my inner Old Prospector.

**The Rainy River Gold Project as seen through the eyes of a Rank Amateur
(Investors take notes, smoke 'em of you got 'em.)**
The first gold bar will be poured in 2017. Not too many months ago the common-room chatter was still a bit negative – "It will never happen," was the buzz. Try telling that to the young lady who is training as an operator of a huge rock hauler. (She won't hear you anyway. She's up there 15 feet in her cab/office.)

So the mine will be mined. The only thing that may hold things up is a crash in gold prices, and with the estimated cost per ounce of production, it would have to be quite a crash.

Let's look at some history now. Ralph Michener or Bob Marvin would tell you how the gold got here in the first place. We will deal with the last hundred years or so.

Bob LeBlanc's grandfather homesteaded in Richardson/Sifton in 1910. The family had been gravitating north from Louisiana for years – there are still some LeBlancs in Lafayette County.

Bob's great-great-(?) uncle knocked some rock off an outcrop in 1914 or thereabouts and saw gold, but he didn't hang around to explore further. The Yukon beckoned, and like any true prospector he headed west/northwest.

Over the years holes were poked into our Shield fringe by Min-Gold (30's or 40's?) and Strat-Mat and Inco (in the50's or '60's.) Core was pulled in Richardson, Sifton and Tait. There were indications of gold, but with the price capped at $35/oz it was of no economic interest.

Others were also interested in our valley. Noranda flew an airborne survey and passed the maps to Mining Corp, a subsidiary. Gerry Daigle came to town to do some follow-up ground work and was scared off by my wood tick story in '66 or '67. I myself chased a couple of anomalies for Noranda west of Off Lake on ground now held by New Gold. Joe Hodge had already come in with Prospector's Airways Syndicate, met Lila Armstrong, opted for civilization at Northwest Bay Lodge and was followed by his brother after Jack had cashed in on the NBU deal.

Then gold was allowed to float, and entering stage left was my old friend Don MacEachern. He and his partner Keith Allen followed up on an Ont. Dept. of Mines till sampling report.

They brought in Doug Hume of Nuinsco, and like a coffee pot in a drill shack, things kept percolating to the point where we now have a gold mine and full employment in the valley. Even the Alberta Tar Sands are envious. New Gold estimates a mine-life of 12 – 14 years. I think they are being conservative (as a reputable company should be.) The thing is – the resource is still open at depth and this is not merely a gold mine – it is a gold zone, stretching off to the west-southwest. Other townships may yet come into play – Nelles, Blue, Worthington (?) and even into Northern Minnesota. Never say Never.

If you are better on the computer than I, you may be able to bring up a satellite view of Lake of the Woods. You will see how the outcrops, valleys and lakes surround the big lake from Shoal Lake on the northwest around the east side and continuing to the southwest. It looks like ripples caused by dropping a stone (Big Island) into a puddle.

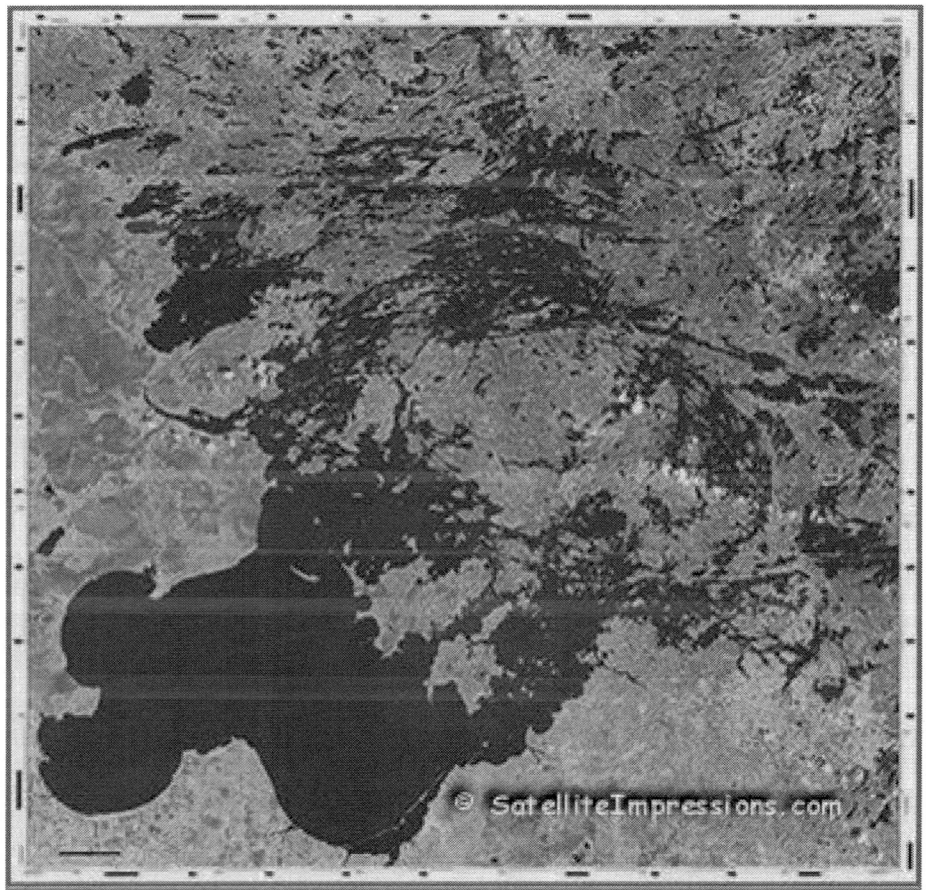

There have been shafts sunk over the years. The old Machin gold mine west of Kenora, for instance. Maybrun took a ton or two of zinc out east of Sioux Narrows, and Arrowsmith has copper in the same area. Chalice may yet open-pit the Cameron Lake deposit and who knows what may transpire at the old Straw Lake Beach mine north of Pipestone. (The real Straw Lake Beach story is worthy of a short chapter on its own.)

But we won't write it. We are far too busy. "Why?" you may ask?

Well, a couple of years ago Bob Marvin asked me how I got to be such a fiddlefoot – so he is to blame. Even as Book Two goes to the publisher we are working on "Book Three – Fiddlefoot Training," completing the trilogy.

Will it ever end? Probably not as long as John Smits (glossary) keeps cranking out 20-dollar bills in his woodshed at Stratton. So we will just hit "Pause" for now.

Epilogue

My co-author, who has had short stories published and who has worked as a reporter/photographer, tells me that all books need an epilogue. But I do not agree. "Epilogue" denotes finality, and why quit when you're having fun. I won this one.

So this will be just another chapter titled:

Chapter XV - How to Write a Book without Really Meaning To

It does not take much to write a book, as far as I see it. One needs a fairly good memory, terrible penmanship, and a brother with supreme computer skills, a decent vocabulary, and who knows a thing, or two, about, commas. It was an accident waiting to happen.

It all started about 2001. I was working seasonally on Lake of the Woods, and with time on my hands in the winter I started to write long letters to my daughter back east. They were background stories – family history, growing up – stuff like that. Unfortunately, she couldn't read my writing. Numerous phone calls were necessary to translate the hieroglyphics.

In 2005 the wife and I lucked out and retired to a Granny Cottage next door to her daughter.

So now I have more time to waste, my quill is still leaving ink blots – and what to do. I call my brother – can I dictate to him, and can he e-mail my daughter? (I don't e-mail, I play solitaire.)

He agrees, as he is also at loose ends, and that winter we do a bit of telephone tag and run out of history. So, remembering my brother's success at writing short stories, I tell him a few anecdotes from my past. He puts them into his computer and one day he says, "I think you have a book here." And that's how it all started.

It's been a long, enjoyable haul. Countless revisions and edits, and you'll still find some mistakes – we are not pros.

But we are hooked – only Wendigo can stop us now!

APPENDIX

Geophysical Assistant Course — Week One

Date	Times	Course Content	Hours
January 10		Students arriving throughout day. Filled in course questionaire and discussion with Frank Cornell re department. General course history.	2
January 11	Morning	Set up classroom, discussed course layout, apportioned equipment. Set up some maps. Discussed different maps and air-photos, proper axe-sharpening. Selected snowshoes and harness.	4
	Afternoon	Compass familiarization, declination and compass use. Transferral of compass heading from map to compass.	2
January 12	Morning	Introduction to staking, sketch of post showing dimensions and marking and tagging. Sample of completed recording sketch. Step by step staking of imaginary Claim in Bridges Twp.	3½
	Afternoon	Make up suitable recording sketch and location sketch for recording imaginary claim. Homework—Picking his own spot, student to describe step by step staking and record same (one claim).	3½
January 13		-staking review and additional points -explanation of different map scales	1 1
		-equipment supply lists (camp gear and groceries) Projected six week stay in bush. Students ordered two weeks grub, pre-ordered next two weeks and drew up final two week order to go out when second order came in. MAXIMUM SUPPLY EFFICIENCY WITH MINIMUM AIRCRAFT EXPENSE. -also covered additions and deletions to camp gear and supplies for summer	4
		-tent site selection and set-up.	1
January 14 -50		-Gene Clyde and Mrs. Proudlock visited. PAYDAY. Discussed clothing problems. -Lands and Forests lecture, on fire safety, camping requirements re clean-up etc. boat and canoe safety and land disposal (crown) (slides & film). Day cut short due to extreme cold in classroom.	2
			3
		TOTAL	28

Geophysical Assistant Course   Week Two

Date	Times	Course Content	Hours
January 17 Mild, Snow	8:30 - 10:00	Review of week one.	1½
	10:00- 11:00	Winter personal requirements in field.	1
	12:00- 5:00	Field trip to Kenora to outfit for personal gear.	5
	8:00 - 10:00	First Aid lecture in Vermilion Bay	2
January 18 Mild	9:00 - 9:30	Review of complete summer & winter field requirements	½
	9:30 -12:30	Review, discussion and note-taking of Lands & Forests lecture.	5½
	1:00 - 3:30	Layout of six claims to be staked in Bridges Turp for course project.	1
	3:30 - 4:30	Class applied for work permit in Bridges Turp. HOMEWORK	
	4:30 - 5:00	Make up proper claim sketch for project staking of six claims.	½
January 19	8:30 -12:30	Practice compass, shooting, snowshoeing, pacing & blazing.	4
	1:30 - 4:30	Covered staking points and discussed next days field staking	3
January 20 Clear, Cold	8:30 -12:30	Completed class on winter and summer tent set-up. Passed out literature supplied by INCO and compared their grub list with ours.	7
	1:30 - 4:30		
January 21	9:00 -11:00	Completion of summer and winter camp requirements.	2
	11:00 -12:30	Radio aerial set-up and proper procedures	1½
	1:30 - 4:30	Class on witness posts. (staking)	3
		TOTAL	37½

237

Geophysical Assistant Course   Week Three

Date	Times	Content	Hours
January 24 Cold, Snow	8:30- 3:30	Four students staked two sides of project group.	7
	8:00-10:00P.M.	First Aid - Vermilion Bay	2
January 25 -42°, clear	8:30-10:00	Additional witness post class work	1½
	10:00-12:30	Line cutting lecture. Set-up of Base Line	
	1:30- 4:30	Starting point, coordinate explanation, cutting and chaining base line. Turning angles with angle board.	5½
January 26 -40°, clear	9:00-12:30	Continuation of line cutting lecture (crosslines) basic mapping, explanation of terms used in line cutting	3½
	1:30 -2:30	Gene Clyde and Frank in to dicuss course..	1
	2:30- 4:00	Discussing individual grading.	1½
	4:00- 5:00	Layout of next days staking.	1
January 27 -25°, Clear	8:30- 1:30	Staking on six claim project group	5
	2:30- 3:30	Discussion of staking.	1
	3:30- 5:00	Small motor trouble-shooting (classroom).	1½
January 28 -18°, Clear	8:30- 1:00	Completed staking on six claim project group review of project staking, distribution of additional material and discussion. Preliminary planning of class	
	2:30- 4:30	individual project staking.	2
		TOTAL	37

238

Geophysical Assistant Course — Week Four

Date	Times	Content	Hours
January 31 Mild,Light Snow	8:30– 9:30	Layout of individual staking project	1
	9:30– 2:00	Individual project staking. 2 groups, 3 men each.	4½
	3:00– 4:00	Recording forms and location sketch for above	1
	4:00– 5:00	First Aid review.	1
	8:00–10:00	First Aid – Vermilion Bay	2
February 1 Mild,Light Snow	8:30– 4:00	Set up, cut and chained base line 0 – 20 W	7½
February 2 Clear,Cold	8:30– 1:30	To Recording Office, Kenora. Record project staking and file abandonment of sqme. Lecture and discussion	5
	2:30– 4:30	Harry Bell, Mining Recorder. Discussion of Bell's lecture, covered main differences re Man. Sask. and Ont. staking.	2
February 3 Clear,Cold	8:30– 4:00	Cutting line and project group.	7½
February 4 Clear,Cold	8:30– 2:00	Cutting line, turned all crosslines off with board and started line cutting	5½
	2:00– 4:00	Class–review and general discussion of line cutting.	2
TOTAL			39

Geophysical Assistant Course Week Five

Date	Times	Content	Hours
February 7 -15°, Clear	9:00- 4:00	Cutting lines on project grid.	7
	8:00-10:00	First Aid	2
February 8	8:30- 4:30	Cutting and chaining lines on project grid.	8
February 9	9:00-12:00	Carl Walston, claims inspector, O.P.M. took group on inspection of one block of claims done by students as individual project.	3
	1:30- 4:30	Wrap-up of line-cutting in class. Covered some problems that might arise and solutions. Covered off-set base lines and topography pick-up.	3
	4:30- 5:00	Magnetometer familiarization.	½
February 10	8:30-12:30	Complete field line cutting and chaining on project group.	4
	1:30- 5:00	Magnetometer - care, operation, some basic theory, note keeping.	3½
February 11	8:30- 3:30	Magnetometer. Set-up base stn. Read base line to establish tie-ins. Practise reading.	4
		TOTAL	35

240

Geophysical Assistant Course   Week Six

Date	Times	Content	Hours
February 14	9:00–10:30	Set up RADEM FIELDSTRENGTH Background and base-stn.	1½
	10:30–12:30	Discussion of base-stn. and base line readings.	2
	1:30– 4:00	How they are used to close off loops	2½
	8:00–10:00	Discussion of Mag theory drawing up line cutting sketch First Aid	2
February 15	8:30–12:30	Started reading base line. Batt. dead on mag.	4
	1:30– 4:00	Make up mag for plotting readings. Discussion of RADEM and some basic theory of primary & secondary fields.	2½
February 16	8:30–12:30	Class on Radem & JEM maps, legends and plotting methods. Some more theory discussion. All maps ready to plot.	4
	1:30–4:30	Read mag base line and Radem base line.	3
February 17	8:30– 1:30	Three groups of 2 men each on mag. radem and JEM practise reading.	5
	2:30– 4:00	plotting of results	1½
February 18	8:30–12:00	Practise reading	3½
	1:00– 2:30	Plotting or results	1½
		TOTAL	33

Geophysical Assistant Course   Week Seven

Date	Times	Content	Hours
February 21	8:30–12:30 1:30– 5:00 8:00–10:00	Practice reading on grid First Aid	7½
February 22	8:30–12:00 1:00–5:00	Plotting first practice maps Practice reading on grid (changed instruments)	3½ 4
February 23	9:00–12:00 1:00– 5:00	Practice reading on grid	7
February 24	8:30–12:30 1:30– 5:00	Practice reading on grid.	7½
February 25	8:30–12:30 1:30–3:30	Practice reading on grid. Mapping.	4 2
		TOTAL	37½

Geophysical Assistant Course   Week Eight

Date	Times	Content	Hours
February 28	9:00– 4:00	Practice reading on grid	7
	8:00–10:00	First Aid	2
February 29	9:00– 4:00	Practice reading on grid.	7
March 1	8:30–12:30	Practice reading on grid.	4
	1:30– 5:00	Mapping	3½
March 2	8:30–12:30	Practice reading on grid	4
	1:30– 4:30	Practice reading on grid	3
March 3	8:30–12:30	Practice reading on grid	4
	1:30–3:30	Mapping	
TOTAL			36½

Geophysical Assistant Course   Week 9

Date	Time	Content	Hours
March 6	8:00-12:00	heavy blizzard, one student in camp, practising mapping (mag contouring). Afternoon off. Decision to close course down this week.	4
March 7	9:00-12:00 1:00- 4:00	Two students in class, discussion of course with students re coverage missed. Some review.	6
March 8	9:00-12:00 1:00- 4:00	Review of pertinent points felt necessary by two remaining students	6
March 9	9:00-12:00 1:00- 4:00	Packing up equipment, personal gear, finalizing students notes & literature	6
March 10	10:00-12:00	Closing down course.	2
		TOTAL	24

COURSE TERMINATES AS OF MARCH 10, 1972.

CLASS/FIELD BREAKDOWN OF HOURS

	CLASS	FIELD
1. General: first aid, camping, general maps, lectures, review	84	0
2. Staking	20½	30
3. Line Cutting	8½	38½
4. Geophysics: Magnetometer, Radem, JEM	29	72
TOTALS	142	140½

TOTAL HOURS CLASS & FIELD   282½

Average Hours per Week   31.4
Average Hours per Day   6.3

CLASS/FIELD HOUR BREAKDOWN

	Class	Field
1. General: first aid, camping, maps, etc. lectures	54	0
2. Staking	20½	30
3. Line Cutting	8½	38½
TOTALS	83	63½

Approximately eight hours spent on mag in class to date.
Approximately four hours to spend on line cutting sketch.

NB. Class/field ratio will swing heavily to field side for geophysics.

Geophysical Assistant Course – Attendance

JANUARY

STUDENT	10	11	12	13	14	17	18	19	20	21	24	25	26	27	28	31
R. Greene	X	X	X	X	X	X	U	E	X	X	U	E	X	X	X	U
R. Kakeway	X	X	X	X	X	U	U	X	X	X	X	X	X	X	X	X
W. Pelly	X	X	X	X	X	E	E	X	X	X	U	U	X	X	X	X
L. Redsky	X	X	X	X	X	X	X	X	X	X	U	X	X	X	X	X
L. Simard	X	X	X	X	X	X	X	X	X	X	X	X	X	X	X	X
D. Tangway	X	X	X	X	X	X	X	X	X	X	U	U	U	U	U	U
J. Tangway	X	X	X	X	X	X	X	X	X	X	X	X	X	X	X	X
G. Wahpay	Late Start					X	X	X	X	E	X	X	X	X	X	E
F. Kahkeeway						Late Start					U					U

FEBRUARY

STUDENT	1	2	3	4	7	8	9	10	11	14	15	16	17	18	21	22	23	24	25	28	29
R. Greene	U	U	U	U	Off Course																
R. Kakeway	X	X	X	X	X	X	U	X	X	X	X	X	X	X	X	X	X	X	X	X	X
W. Pelly	U	X	U	U	E	U	U	U	U	X	X	U	X	X	X	X	X	U	U	U	U
L. Redsky	X	X	X	X	E	X	X	X	X	X	X	X	X	X	X	X	X	X	X	U	U
L. Simard	X	X	U	X	X	X	X	X	X	X	X	X	X	X	X	X	X	X	X	X	X
D. Tangway	U	U	Off Course																		
J. Tangway	X	X	X	X	X	X	X	E	E	X	U	U	U	U	X	X	X	X	X	U	U
G. Wahpay	X	X	X	X	X	X	X	X	X	U	E	X	X	U	E	X	X	X	X	E	X
F. Kahkeeway	U	X	X	X	X	X	X	X	X	U			Off Course								

MARCH

STUDENT	1	2	3	6	7	8	9	10
R. Kakeway	X	X	X	U	U	U	U	U
W. Pelly	U	X	X	U	U	U	U	U
L. Redsky	U	U	X	U	U	U	U	U
L. Simard	X	X	U	X	X	X	X	X
J. Tangway	U	U	U	U	U	U	U	U
G. Wahpay	X	X	X	E	X	X	X	X

X-Present
U-Unexcusable Absence
E-Excusable Absence

Glossary

(Most of these terms pertain to book one, but are worth repeating here.)

Ambroid: A tube of material which hardens when exposed to air: used to repair cedar strip canoes or any canvass-covering.

Azimuth: Compass direction, usually in terms of degrees of the compass, i.e.: 30 degrees east of true north

Bombardier: (en Francais, Bombardier Auto-Neige) a Quebec company, manufactured two enclosed types of snow vehicles. They made two sizes, narrow track and wide track. A Chrysler flat head six-cylinder motor, located in the rear of the unit, powered them. A three-speed transmission coupled to a high-numbered (low ratio) differential delivered power to sprockets at the rear of a rubber belted track on each side. These rubber belts had cross-cleats, which rolled on solid rubber tires. At the front were two skis on knee action struts to steer the thing with, although the turning circle was anything but tight. When running on a previously broken bush trail the skis followed the track fairly well, but if you were breaking a fresh track the sharper corners required a lot of back-and-forth. Two bucket seats up front held the driver and one passenger, the rest of us sat on two canvass upholstered benches on each side over the track tunnels, with another upholstered bench across the rear in front of the enclosed engine. It would seat eight men comfortably. On cold days we could open a hatch on the engine compartment to get more heat. The centre floor had lots of room for gear or cargo. Top speed was 30 mph, but bush road conditions seldom allowed more than 15 to 20 mph. It was pretty darn comfortable, really, unless you found it necessary to use a road frequented by Muskeg tractors – these roads were more of the roller coaster variety.

Canadian Nickel had seven of these units. Numbers 2 to 6 were used regularly, #7 was a standby unit. Number 1 was Hector's personal machine. Hector was the road commissioner/maintenance chief. I won't say much about him, but the fact that he felt he was entitled to his own personal machine says it all.

Bore Hole: Diamond drill hole

Bushwhacker: Often used as a pejorative, for all those who make their living slogging through muskeg while fighting off black flies and mosquitoes.

Canadian Nickel: Inco's Manitoba exploration wing.

Candled Ice: In the spring the melting ice will get vertical holes in it, looking like closely packed candles: also known as rotten ice. Stay Off!

Chalcopyrite: A copper/iron sulfide mineral.

Churchill Line: CN built this spur line to the newly created port of Churchill, Manitoba in the '50's, now owned by Omnitrax.

248

Cookee: The cook's helper and all around good guy, in charge of washing dishes, carrying water, cleaning counters, sweeping floors and carrying garbage.

Contact: The joint-plane where two different formations meet. Ore bodies frequently occur along contacts.

CNR/CN/Canadian National Railways: Once Canada's National rail line, now owned primarily by Bill Gates.

Dike: Any mass of igneous rock, which, in a state of fusion, has entered a fissure in other rocks and has chilled and solidified. Dikes are usually vertical.

DOT: Federal Department of Transport: among other duties. The DOT maintains national regulations and standards for the flying part of Canadian Life.

Float: Loose scattered rocks, not easily identified as to source: If a prospector can establish glacial flow direction, he may be able to find the source - the smoother the float, the farther it has traveled. When Grandpa was picking stones in the barley field, he was harvesting float.

Fold/folding: A roll or bending in rocks or veins.

Gossan: The iron-bearing deposit filling the upper parts of veins or covering masses of pyrite. Oxidization of the iron component of the deposit lends it a rusty colour.

John Smits: Bless his heart, he bought a copy of Book One. Then he wanted his $20 back. He enjoyed the book but had finished reading it – he would trade it for the twenty. I told him I could not do it – I no longer had the $20 bill. I had laundered it along with all the Stratton 20s – they all had identical serial numbers.

Muskeg Tractor: This was also built by Bombardier. It had wider tracks than the snow machine (generally referred to in this text as the Bombardier), with two rows of bogeys on each side, and was steered by hand clutches like a caterpillar tractor. A cargo deck on each side sat over the track tunnels. Inco used the J-5: The driver sat low in the centre, between the two cargo decks and in front of the motor, and a canopy swung back to access the "cockpit." It was geared lower than the Bombardier – top speed 15 mph. It was used extensively by Midwest and other drilling companies winter and summer. It had good flotation on muskeg, hence the name.

These machines – snow and muskeg – were the mainstay of the exploration camps. The snow machine is still widely used by commercial fishermen on the lake ice and you'll also find them in the tourist industry. The Skidoo was still being refined and I would not see them in camp until the winter of '62 in Northern Quebec.

NFB: National Film Board of Canada – funded by tax dollars.

Packsack Diamond Drill: Developed and manufactured by Reg Minogue in North Bay, Ontario, it was meant for shallow drilling to test anomalies or surface showings. It was basically a 7 hp chain saw motor with the chain sprocket replaced by a heat-treated coupling to drive the drill string - five-foot sections of 1 1/4 inch hollow steel rod. A five-foot core barrel, reaming shell and diamond set drill bit, preceded the rods down the hole. A swivel tee between the motor and drill string allowed for water to be pumped down the hollow rods to the drill bit, to flush away rock cuttings and bring them up the outside of the hole to the surface, as well as providing cooling and lubrication for the process. The diamond faced hollow drill bit and reaming shell kept the hole constant and made a rock core, which would fill the five-foot core barrel. When the core barrel was full, the crew would pull the drill string by hand, remove the 7/8" rock core and store those from the target area in core boxes. (This is basically the same principal used by the larger drills, the only real difference being the larger drill string is pulled by a winch.)

A "Johnson Bar" was used to apply pressure to the bit while drilling. It was simply a steel bar five feet long with a yoke on the drill, attached by chain to a hook fixed to a convenient stump or rock bolt near the hole collar. One man held the drill motor steady while another applied downward pressure on the long end of the Johnson Bar. As the hole progressed, more sections of drill rod were added.

Casing, steel tubing slightly larger than the drill string, also with a diamond encrusted "shoe," would first be driven through any overburden to bedrock to seal the hole, keeping gavel, dirt and small rocks from falling in.

Supposedly good for 200 feet, it was seldom used past 100. It was heavy to pull, and lack of power to overcome downhole friction limited the depth. In extremely hard rock such as quartzite, it was difficult to keep enough down pressure to avoid polishing the diamonds: Then the string would have to be pulled, and the bit replaced with a fresh one, the old one being sent back to the manufacturer to have the diamonds reset with fresh, sharp points protruding.

Deep overburden and boulders could also negate the Packsack drill's effectiveness, once again because of lack of power.

The drill was hard on the motorman, a position that was often traded off during the workday. Your hands would continue to tingle long after the supper dishes were done.

By 1964 the Winkie drill would replace the Packsack drill. But The Packsack drill did its job well within limits and was a good prospecting tool. Some are still to be found in use by low-budget prospectors.

Spotting the Borehole: A picket would be planted at the hole location. On it one would write the azimuth and the angle of the hole to be drilled.

Strike: General direction of the rock formation in an area.

Step: Aircraft floats are made with a break in the bottom surface about half-way down, with the rear bottom surface about two inches higher than the front. As the plane picks up speed and some lift is generated, the higher rear surface lifts out of the water, reducing the amount of drag and surface tension so the plane can accelerate to flying speed.

STOL: Short Takeof and Landing

Swash Plate: The plate fastened to the rotor hub on a helicopter that controls the angle of attack of the blades, increasing it on the backward sweep and decreasing it on the forward sweep of the blades' travel during forward motion. This gives uniform lift on both sides of the craft. It is a major wear area, and is also very critical to the safe operation of the craft. In cold weather the wear and stress can easily cause failure, hence rotorcraft such as the Bell G series had lower limits on the temperature they could be flown in.

Test Tube Etching: A mild acid/water solution is inserted in a glass tube which is lowered into the drill hole and left for a short period; the acid etches the glass, and when inspected shows the angle of the hole from vertical.

Winkie drill in Action (One picture is worth a thousand words)

Winter Road: Built after freeze-up and hopping from lake to lake, these were the vital supply lines to the north for many years. They are still used in some places where rivers have to be crossed and funds are not available for bridges. (Fort Chipewyan, Sk. and Yellowknife, NWT for example.)

Wobble/wobbling: an unscheduled, unplanned and usually illegal work stoppage, most often of a unionized work force.

Made in the USA
Lexington, KY
10 July 2019